War by What Means, According to Whose Rules?

The Challenge for Democracies Facing Asymmetric Conflicts: Proceedings of a RAND–Israel Democracy Institute Workshop, December 3–4, 2014

Amichay Ayalon, Brian Michael Jenkins

For more information on this publication, visit www.rand.org/t/CF334

Library of Congress Cataloging-in-Publication Data is available for this publication.
ISBN: 978-0-8330-9169-7

www.rand.org

Preface

The workshop summarized here had its origins in a series of conversations in California and Israel between Admiral Amichay Ayalon and Brian Michael Jenkins—a true meeting of the minds. Both men had spent years dealing with the challenge of insurgent warfare and terrorism. Having years of military experience, neither was philosophically opposed to the use of military force when necessary, but both found purely military approaches to be inadequate. Both men were dedicated patriots but unashamed to be self-critical. Both men had been around too long to care about what might be politically advantageous or prudent to say—or not say. For years, both had worried that the continuing terrorist threat and the accumulating security measures implemented to combat it were insidiously, but profoundly, changing society. At issue was not simply whether democratic societies could defeat irregular adversaries on the field of battle or suppress terrorist groups at home, but whether liberal democracies could match the evolving capabilities of those whose belief systems allowed them to employ tactics that defied the laws of war, deliberately target noncombatants, and sacrifice their own civilian populations and yet remain liberal democracies.

Few nations have had more experience in dealing with this threat than the United States and Israel, though not all of that experience was successful. Both nations are tempestuous democracies. Both nations field highly trained military forces equipped with the latest technologies. In strict military terms, Israel could dominate its foes. The United States remains the world's superpower. Yet both find themselves drawn into lengthy, costly, and frustrating conflicts that are seen as essential but soon become unpopular. Both nations face strong criticism in the international community, sometimes simply because of their military prowess and superiority. Some of this criticism is genuine and, at times, deserved, but some of it reflects deeply rooted anti-American and anti-Israeli sentiments. That observation, however, is no excuse for not trying to do things better.

The idea of a collaborative effort emerged from the conversations. This workshop was seen as the first step. It would provide an opportunity for experienced analysts from Israel and the United States, many with long backgrounds in military service, intelligence, diplomacy, and research, to exchange ideas, free of bureaucratic constraints or political agendas. Its realization was made possible by the generous contribution of David A. Lubarsky.

The assembled group met for two days in December 2014. On the second day, a number of U.S. government officials were invited to join the sessions to hear firsthand the directions the discussions were taking and to lend their comments and advice. The event culminated in a public forum at the National Press Club in Washington, D.C., where James Kitfield, currently a senior fellow at the Center for the Study of the Presidency and Congress, acted as moderator of a discussion with Ayalon and Jenkins.

The publication of these proceedings is the second step. It is hoped that they will provoke further comments and ideas. The next step is to be determined. The dialogue will continue.

The workshop was conducted within the Center for Middle East Public Policy, part of International Programs at the RAND Corporation. The center brings together analytic excellence and regional expertise from across the RAND Corporation to address the most critical political, social, and economic challenges facing the Middle East today. For more information about the RAND Center for Middle East Public Policy, visit www.rand.org/cmepp or contact the center director (contact information is provided on the center's web page).

Contents

Figures

Acknowledgments

An international gathering such as this workshop requires a collective effort. There are many people to thank. First, the workshop would not have been possible without the generous support of David Lubarsky, who, from the start, remained a determined champion and supporter of the collaborative effort. The RAND Corporation and Israel Democracy Institute invested additional resources to support the meeting.

Dalia Dassa Kaye, director of RAND's Center for Middle East Policy, provided intellectual leadership and practical guidance. RAND's administrative and support personnel ensured smooth hosting of the event. We are especially grateful for the hard work of Francisco Walter, Saci Detamore, and Carolyn Higgins.

In accordance with the terms of our invitation, we cannot acknowledge by name the U.S. government officials who attended the second day of the workshop, but we wish to note their participation and helpful comments.

None of our colleagues at RAND or the Israel Democracy Institute was fully compensated for the time invested in preparing for and attending the workshop. They participated out of conviction that the topic is of vital importance and analytical collaboration is useful. We thank them for their contribution.

The proceedings presented here benefited from the skillful editing of Janet DeLand.

Donor Appreciation

Funding for this project was provided by a gift from David A. Lubarsky, in memory and appreciation of his parents, Martin and Lorna Lubarsky. The donor also wishes to recognize the tireless support and collective efforts of Brian Michael Jenkins and Amichay Ayalon, both of whom were essential for this project.

Abbreviations

CIA	Central Intelligence Agency
IDF	Israel Defense Forces
IDI	Israel Democracy Institute
ISIS	Islamic State of Iraq and Syria
NATO	North Atlantic Treaty Organization

CHAPTER ONE
Introduction

Brian Michael Jenkins

In 2014, the Israel Democracy Institute (IDI) and the RAND Corporation initiated a collaborative research effort aimed at developing new strategies to cope with asymmetric conflict in all of its dimensions, including military operations, human rights and the role of law as it affects conflict, media, public opinion and political warfare, international diplomacy, the internal politics that come with democracy, and the preservation of civil liberties. The objective of this effort is the creation of an analytical framework, doctrines, and strategies that will enable democracies to effectively defend themselves against asymmetric threats while maintaining their commitment to democratic principles and humanitarian values.

As conventional interstate warfare has declined, modern democratic states increasingly find themselves engaged in asymmetric conflicts against nonstate actors that often employ terrorist tactics, which not only threaten the security of the state but also provoke reactions that threaten the fundamental values that define democracy. Asymmetric conflicts include state-sponsored attacks using irregular forces and unconventional means to apply pressure on their adversaries. This pattern of conflict seems likely to prevail for the foreseeable future.

As in any democracy, sharp differences have arisen among lawmakers, intelligence officials, military leaders, security professionals, legal scholars, and ethics advocates on these difficult issues: What domestic security measures are tolerable in a democracy? What rules should govern the collection of domestic intelligence, especially in light of new technologies that equip authorities with unprecedented capabilities for surveillance? How can conflicts that lack a definition of victory or even clear beginnings and ends be brought to a close, or are the United States and Israel condemned to open-ended wars? When conflicts persist without resolution, how can democracies prevent government from accumulating extraordinary powers to the point that those powers permanently alter the relationship between the government and the governed?

In conflicts where there are no territorial boundaries, when and where is the use of military force abroad justified? How does one even define the zone of conflict? Are pre-emptive attacks legitimate? When force is applied, what weapons may be used with what rules of engagement? What is the status of those captured or arrested? Are asymmetric conflicts governed by their own laws of war and bodies of ethics, different from those of routine law enforce-

> *How can conflicts that lack a definition of victory or even clear beginnings and ends be brought to a close, or are the United States and Israel condemned to open-ended wars?*

1

ment or other traditional modes of armed conflict? Should they be governed by such laws and ethics?

Few nations have accumulated as much experience in dealing with asymmetric conflicts as Israel and the United States. For the past half-century, the two countries have been almost continuously engaged in asymmetric conflicts with homeland and national security dimensions—domestic and international terrorist campaigns and military engagements abroad with guerrillas and insurgents—although the experiences of the two nations differ greatly. While both nations have, for the most part, succeeded in protecting their populations (the obvious exceptions being the 9/11 attacks in the United States and the campaign of suicide bombings in Israel early in the past decade), albeit at great cost, criticism from abroad and at home have questioned their methods. It is time for a fundamental review, not an audit to assign blame for perceived errors or excesses, but an effort to distill hard-learned lessons and explore new approaches.

As a first step, representatives from the RAND Corporation and IDI organized a two-day workshop, which was held in Washington, D.C., on December 3 and 4, 2014. The workshop joined two teams of researchers, one from Israel led by Admiral Amichay Ayalon, the other comprising RAND analysts led by Brian Michael Jenkins. Workshop attendees included researchers from a variety of fields, including law, history, media and communications, terrorism, and military tactics and strategy.

Initial workshop issues and analyses were then shared with invited U.S. government representatives from the National Security Council, the Department of State, the Department of Defense, and the Intelligence Community. The workshop culminated with a public forum moderated by James Kitfield, a senior fellow at the Center for the Study of the Presidency and Congress, at the National Press Club in Washington, D.C., on the evening of December 4, 2014, in which Ayalon and Jenkins shared some of the workshop's findings and recommendations. The introduction to this public forum was made by Yohanan Plesner, president of IDI.

The proceedings presented here summarize the two days of discussion. The discussion began, of course, with welcoming remarks that underscored the wide range and complexity of on- and—increasingly—off-the-battlefield issues that are part of contemporary conflict, the necessity of candid dialogue rather than the defense of established positions, and the objective of formulating the right questions instead of jumping to conventional answers.

Underlying these remarks—indeed, underlying the entire discussion—was a sense of frustration that military superiority, even military success, no longer counted as it did in past conflicts. This was accompanied by a healthy humility about being able to find the correct formulas for success and concern that asymmetric conflicts were having a pernicious effect on the democracies being defended, obliging them, luring them toward increasingly oppressive measures.

CHAPTER TWO

Welcomes

Dalia Dassa Kaye, director of the RAND Center for Middle East Public Policy, welcomed attendees to the workshop, which she described as a beginning of a process in which ideas would be freely exchanged and topics for future research identified. She stated that she was delighted to have IDI as RAND's partner in the workshop and thanked Ayalon for assembling his team of Israeli researchers and academics from across Israel. In her view, the workshop was a test case for possible similar workshops with other Israeli institutions.

Ayalon added his welcome on behalf of the Israeli delegation, joking that he saw the workshop as a "boot camp"—an intensive program organized by a retired Israeli admiral and former director of the Shin Bet (himself) and a former Special Forces officer (Jenkins). He stressed that the growing asymmetrical character of warfare is a new phenomenon, one that is especially important for him, as an Israeli. He hoped to learn from the experiences of other nations. While answers were welcome, this workshop would focus on formulating the right questions.

In his welcoming remarks, Jenkins observed that discussions about the workshop had begun more than a year earlier with the simple idea of gathering experienced researchers from Israel and RAND into one room to begin what he hoped would be a spirited discussion addressing a wide array of topics. Jenkins noted that, from the beginning, he and Ayalon saw the meeting as a *workshop*, not a conference where formal papers would be presented. Its objective was to distill questions and hypotheses that would guide future inquiries. Like Kaye and Ayalon, Jenkins expressed his hope that the meeting would be the first step toward continuing collaboration between RAND and IDI, along with the Interdisciplinary Center Herzliya, Haifa University, and other Israeli institutions.

After further introductions, the workshop got down to business. Discussions were organized into seven sessions. Each of the first six would address a specific aspect of contemporary asymmetric conflict. The seventh would attempt to summarize some of the key observations and sketch out areas for future inquiry.

Each session began with opening remarks—not formal papers, but observations and reflections, some of which reflected previous, ongoing, or future research. These were intended to provoke further debate. In each session, the opening remarks were followed by comments from a designated discussant. This led to open dialogue.

Modified Chatham House rules were invoked. In these proceedings, we have identified the authors of the opening remarks, the commentators, and, in some cases, individuals referring to specific research they have done, but we have not connected all of the remarks to specific participants in order to adhere to Chatham House rules and to encourage spontaneity and candor in our discussions.

Where a participant was calling attention to a specific piece of published research he or she had done, we identified the speaker with his or her permission. We also circulated the draft of the proceedings and offered participants the opportunity to make corrections and reinsert their names at appropriate places, if they so desired.

Editor's Note

As editor, I must accept responsibility for the quality of the summary of these proceedings. A chronological reproduction of everyone's remarks would read like the transcript of a football match and would make little sense. Each intervention inspires multiple responses and ripostes—a phalanx of name cards spring from the prone to present arms. They must be addressed in order, which means that a comment may follow a half-dozen other comments on something said minutes before.

It made more sense to try to reassemble the remarks around certain themes that arose in each session. Inevitably, that required combining and pruning. No doubt some damage was done to individual interventions, for which I apologize in advance to my fellow participants.

Brian Michael Jenkins

CHAPTER THREE

Session One: The Changing Terrorist Threat and America's Evolving Response

Opening Remarks by Brian Michael Jenkins

Jenkins opened the first session with a broad review of recent military history. The accompanying discussion was led by professor (and former general) Yishai Beer. Jenkins began his remarks by acknowledging a high degree of personal frustration felt by many Americans with the course of events in Afghanistan and Iraq. U.S.-led military interventions in both countries had become the two longest wars in the history of the American republic, and, although not the bloodiest conflicts in American history—a position still held by the Civil War—the two wars had resulted in thousands of Americans dead and tens of thousands wounded, and they had cost trillions of dollars in treasure. The total enemy and civilian casualties of the two wars are estimated to be in the hundreds of thousands.

Despite this investment in warfighting and nation-building, a still precarious security situation was obliging the United States to reconsider its planned withdrawal from Afghanistan. And while all U.S. forces had departed from Iraq, events there and in Syria had brought about the renewal of American bombing and pressure from domestic political quarters to reintroduce American combat forces.

Having recently returned from the Middle East, Jenkins also noted the frustration of Israeli intelligence and military officials with the repeated military engagements with adversaries, such as Hezbollah and Hamas, that produced no permanent effect. Jenkins recounted his exchange with a senior Israeli military officer shortly after the 2014 military operation in Gaza. He had asked the officer what the Israelis had achieved in this latest round of fighting. The officer readily conceded that the operation had cost the Israelis in casualties and international reputation but had led to no clear and immediate resolution of the situation in Gaza—he only hoped that this third round of fighting would produce a cease-fire that lasted longer than the previous two before the next Gaza war broke out.

This led to a fundamental question regarding such developments as the refinement of terrorist tactics over the years; improvements in technology, especially information technology; the growing capabilities and active role of both the traditional media and rapidly evolving social media, not as mere observers but as shapers of perceptions; and changes in

> *[W]hile all U.S. forces had departed from Iraq, events there and in Syria had brought about the renewal of American bombing and pressure from domestic political quarters to reintroduce American combat forces.*

public attitudes toward armed conflict: Have these developments, which coincide with the evolution of international humanitarian law, which sometimes conflicts with the international laws of war, combined to place democracies at a disadvantage when engaged in asymmetrical warfare?

This is not a question about constraints on the application of military force, and the answer ought *not* to be that success against adversaries that operate outside the rules of war requires greater ruthlessness on the United States's and Israel's part. That would mean abandoning the fundamental values that the United States and Israel claim to represent and defend and the moral high ground that the two nations and their publics at home and allies abroad deem important, although the unevenness of international condemnations and partisan domestic politics engender some cynicism about self-righteous critics, fickle public opinion, and feckless politicians. The United States and Israel should be concerned with finding the right strategy or combinations of strategies to deal with specific situations. What works? And what clearly has not worked?

To make the point that we are not talking about some theoretical future war, Jenkins reviewed America's recent military history. Since the end of the Vietnam War, there have been about 20 major "terrorist" events that were serious enough "to get the lights on at the White House in the middle of the night" and raise at least the possibility of military action. (The quotation marks around "terrorist" indicate that some would argue about the designation of all of these as terrorist events.) Almost all of these events had their origins in ongoing conflicts in the Middle East.

The first of these events was the hijacking of three airliners by members of the Popular Front for the Liberation of Palestine who flew the planes to Dawson's Field in Jordan in September 1970. The passengers of the airliners sat in the field, surrounded by armed gunmen while their leaders made their demands at press conferences. In response to the tense hostage situation, President Richard Nixon put the 82nd Airborne Division on alert and the Sixth Fleet out to sea. The hostages were then transferred to Amman, which made an armed rescue impossible. Most of the hostages were freed in a few days; all were eventually released.

Other incidents on Jenkins' list included the armed assault on American passengers at Israel's Lod Airport in May 1972; the seizure and eventual murder of American diplomats at the Saudi embassy in Khartoum in March 1973; the takeover of the American embassy by militants in Tehran in November 1979; the bombings of the American embassy and U.S. Marine Corps barracks in Beirut in 1983; the continuing captivity of American hostages in Lebanon during the 1980s; the hijacking of TWA flight 847 in June 1985; the hijacking of the Italian cruise ship *Achille Lauro* in October 1985; the bombing of TWA flight 840 in April 1986; the attack at the Rome Airport in December 1985; the deaths of American soldiers in the bombing of a Berlin discotheque in April 1986; the sabotage of Pan Am flight 103 in December 1988; the Iraqi plot to assassinate former President George H. W. Bush in April 1993; the bombings of American embassies in Africa in August 1998; the attack on the USS *Cole* in October 2000; the 9/11 terrorist attacks; and most recently, the murder of American diplomats in Benghazi, Libya, in September

[T]he unevenness of international condemnations and partisan domestic politics engender some cynicism about self-righteous critics, fickle public opinion, and feckless politicians.

2012. Seven of these events prompted U.S. military intervention.

> *[D]espite their high failure rate, these guerrilla and terrorist campaigns can last decades, can be exceptionally savage in nature, and can be frustrating for military commanders to confront.*

Putting these interventions in the broader context of American military interventions abroad during the same period, Jenkins noted that there have been 19 overt U.S. combat operations since the end of the Vietnam War. Most of these have occurred in the Middle East, North Africa, or Southwest Asia: Iran (1980), Lebanon (1983), Libya (1986), the Gulf War (1991), Somalia (1993), Iraq (1993 and 1996), Afghanistan (1998), Sudan (1998), Afghanistan (2001), Iraq (2003), Libya (2011), Iraq (2014), and Syria (2014). Only the brief dispatch of troops to Cambodia in 1975 to rescue the crew of the SS *Mayaguez*, the invasion of Grenada in 1983, the invasion of Panama in 1989, the deployment of troops to Bosnia in 1995, and military operations in Serbia in 1999 occurred outside of the Middle East, North Africa, or Southwest Asia.

Jenkins then focused on success rates of various insurgent and terrorist groups, citing the research of his RAND colleagues Seth Jones and Martin Libicki. Jenkins pointed out that, ultimately, these groups also faced frustration. Jones's research identified 51 insurgencies since 1975, 12 of which are ongoing, and 14 cases in which the insurgents were victorious. In the remaining 25 cases, the insurgents' military campaigns had ended in less than victory (Jones and Libicki, 2008). Similar research showed that only 10 percent of the campaigns of 648 terrorist groups active since 1968 could be considered victorious (Connable and Libicki, 2010).

Jenkins pointed out that, despite their high failure rate, these guerrilla and terrorist campaigns can last decades, can be exceptionally savage in nature, and can be frustrating for military commanders to confront. This is not new. For example, at the beginning of the 19th century, Spanish guerrilla tactics were able to pin down 400,000 of Napoleon's troops—Napoleon wrote that the Spanish engagement was his most frustrating military encounter.

Jenkins then outlined a series of questions to frame the workshop discussion:

1. Are terrorists developing new mechanisms that render ineffective the superior military technology of Western governments? Meanwhile, is public opinion, particularly in Western democracies, imposing growing constraints on the application of military force?
2. Are the perceptions of a perpetual terrorist threat, combined with enabling technological developments, pushing democracies toward creating a "security state" in which individual liberties will be incrementally compromised?
3. Have changes in Western attitudes made society so sensitive to each individual casualty as to render the use of military force ineffectual? That may seem to be a positive development, but it has significant policy implications.

Expanding on this final point, Jenkins noted that the apparent increase in sensitivity to individual casualties in democracies paralleled the decline of deaths attributable to war. World Wars I and II caused more than 100 million deaths. The United States suffered 250,000 deaths in World War II. In Vietnam, the American death toll was approximately 58,000. In Iraq, approximately 5,000 Americans died. And 3,000 Americans were killed in Afghanistan.

Similarly, approximately 10,000 Israelis died in the first four Israeli wars (1948–1949, 1956, 1967, and 1973). Since 1973, fewer than 1,000 Israelis have died in military operations in Lebanon, the West Bank, and Gaza.

As citizens of democracies have become acutely sensitive to their own casualties, each death has had an increasingly great effect. At the same time, fear has escalated, elevating foes of limited capability to existential threats. For example, former Secretary of Defense Chuck Hagel, referring to the 2014 military offensive carried out by the Islamic State of Iraq and Syria (ISIS), described ISIS as "an imminent threat to every interest we have," and a four-star general claimed that "World War III is at hand." Conceding that ISIS is a new kind of terrorist adversary, Jenkins characterized these comments as "extraordinary," given that the adversary about to begin World War III and that threatens every interest that the United States and Israel have is not a nuclear-armed Soviet Union or Russia, but a desert force of 30,000 irregular fighters with a few captured tanks and a lot of pickup trucks.

This introduction served as a segue to Jenkins' forthcoming paper, "How Combating Terrorism Became a War: The Evolving Use of Military Force in Response to Terrorism," which traces what Jenkins calls "the long arc" of the military's growing—and often reluctant—involvement in counterterrorism operations. Jenkins argued that, when contemporary terrorism emerged in the 1960s and early 1970s in Latin America, the Middle East, and the West, the United States did not initially seek to fight terrorist groups directly. It simply sought to isolate local conflicts and thus prevent the spillover of violence into the international community in the form of airline hijackings, attacks on diplomats and diplomatic facilities, or other tactics calculated to call attention to their causes, spread fear and alarm, and coerce foreign governments to make political concessions or alter their policies.

Terrorist tactics were regarded as outside of the norms of war, and the United States sought to build an international consensus to outlaw terrorism regardless of its political cause. By building international cooperation, the United States was aiming to deny terrorists sanctuary and ensure prosecution of those who escaped local authorities. That was a State Department job or a task for the Department of Justice. Initially, there was no defined military mission to fight terrorism. In fact, the U.S. military was reluctant to become involved.

The specter of Vietnam continued to haunt American military thinking. From the military's perspective, Vietnam had been not just a failure but a debacle. It had turned the American public not merely against the war but against the military, and it nearly destroyed the institution of the armed forces. The post-Vietnam Pentagon was determined to rebuild the armed forces, while avoiding irregular conflicts and "unwinnable" missions. The counterinsurgency capabilities developed during Vietnam were dismantled to ensure that civilian political leaders could not engage in new Vietnams or other similar adventures.

The U.S. military was determined to engage only in classic conventional-war missions in which superior military

> *As citizens of democracies have become acutely sensitive to their own casualties, each death has an increasingly great effect. At the same time, fear has escalated, elevating foes of limited capability to existential threats.*

power would prevail. From the military's perspective at the time, the real threat to America's national security remained the Soviet Union, not urban guerrillas or terrorist kidnappers. Although it was accepted that, in particular circumstances, military force might have to be used to rescue Americans held hostage—

Shultz argued that unwillingness to apply military force except under the most rigorous preconditions undermined American diplomacy.

as contemplated in 1970, ordered in 1975 to rescue the crew of the SS *Mayaguez*, and attempted unsuccessfully in 1980—these were dangerous sideshows that could tarnish the perception of U.S. military prowess. These attitudes changed only gradually and in response to specific events.

The first elbow in the curve came in 1983, when terrorists drove a massive truck bomb into a building housing U.S. marines in Beirut. The devastating attack, which killed 241 American military personnel, was accompanied by a simultaneous bombing of a building housing French paratroopers, killing 55. The French and Americans initially agreed to conduct a joint strike in retaliation, but the Americans backed out at the last minute.

In reviewing the Marine Corps–barracks bombing, a Pentagon commission led by Admiral Robert Long concluded that terrorist warfare could have a significant political impact—it represented a new mode of warfare for which the U.S. military was inadequately prepared. This set the stage for an interagency fight between Secretary of State George Shultz and Defense Secretary Caspar Weinberger. Shultz argued that unwillingness to apply military force except under the most rigorous preconditions undermined American diplomacy.

The Secretary of State prevailed, and the change in official American policy regarding the approach to terrorism was codified in National Security Decision Directive 138 (Reagan, 1984). American policy henceforward would be to make counterterrorism a military mission. Referring to "war against terrorism," the directive instructed the Secretary of Defense to improve U.S. capabilities to conduct military operations to counter terrorism against U.S. citizens, develop a military strategy to combat state-sponsored terrorism, and develop a full range of military options to combat terrorism.

Nevertheless, the debate continued, with the military remaining resistant to missions other than conventional wars. The Pentagon set forth a set of prerequisites that had to be met prior to deployment of American combat troops. These conditions were initially articulated by Weinberger in November 1984 and were later repeated by General Colin Powell. Subsequently, this military thought was known as either the Weinberger Doctrine or the Powell Doctrine.

The Weinberger Doctrine, articulated in a speech on November 28, 1984, entitled "The Uses of Military Power," stipulated the following:

- The United States should not commit forces to combat unless vital national interests of the United States or its allies were at stake.
- U.S. troops should be committed only with the clear intention of winning.
- U.S. combat troops should be committed only in pursuit of clearly defined political and military objectives and with adequate resources to achieve them.
- The relationship between objectives and size and composition of forces committed should be continually monitored and adjusted.

- U.S. troops should not be committed to battle without a reasonable assurance of support by American public opinion and Congress.
- The commitment of U.S. troops should be considered only as a last resort when all other options have failed ("The Uses of Military Power," 1984).

Although terrorists offered few lucrative targets for conventional military attack, the United States took its first military action in response to state-sponsored terrorism. The United States did not seek to directly engage terrorist groups; instead, it would engage states viewed as sponsors of terrorism, including Syria, Iraq, Iran, Sudan, and Libya. The inclusion of Libya reflected growing concerns about terrorist attacks organized by the Abu Nidal Organization, which was based there.

The terrorist bombing targeting U.S. soldiers at a Berlin nightclub in 1986 prompted the U.S. bombing of Libya. That established a precedent for responding to terrorism with military force, although the United States was reluctant to use the term "military retaliation." In 1993, the United States bombed targets in Iraq in response to an Iraqi-instigated assassination plot against former President George H. W. Bush; and in 1998, it bombed targets in Sudan and Afghanistan in response to the bombings of American embassies in Africa.

While the United States initially tried to combat terrorism by gaining support for international treaties that outlawed specific terrorist tactics, the focus of U.S. efforts gradually changed to attacking terrorist groups directly. The shift began with the practice of identifying terrorist organizations in the State Department's 1987 annual report titled *Patterns of Global Terrorism*, under the heading "Worldwide Overview of Organizations That Engage in Terrorism" (U.S. Department of State, 1988). As this list was for information purposes only, it had no legal standing.

In 1996, however, Congress enacted legislation authorizing (in effect, instructing) the Secretary of State to designate "foreign terrorist organizations" (Pub. L. 104-132, 1996). The first designations were made in 1997. The logic behind the formal designation reflected a growing determination in Washington to take the offensive against terrorist groups, initially by eliminating their funding and other forms of material support. But as is so often the case, the legislative authorization was driven by specific circumstances—in this case, events in Israel. Drafting of the legislation started in 1994, prompted by the Hebron massacre and bus bombings in Israel. As an interim measure, economic sanctions were imposed on 12 Middle Eastern terrorist groups—ten Arab and two Jewish—that threatened violence against the peace process.

Officially designating terrorist groups had broader policy implications. It shifted the focus of U.S. counterterrorism measures from the terrorist *event* to the terrorist *group*. The objective became not merely to outlaw specified terrorist tactics but to go after specified terrorist groups. Although the initial measures were economic, the designation opened the way for more direct engagement of terrorist groups, accompanied by a growing military role.

The events of September 11, 2001, further changed the U.S. course of counterterrorism. The scale of the 9/11 attacks was unprecedented in the annals of terrorism. Propelled by fears of further or even worse large-scale attacks, President George W. Bush declared a "Global War on Terror." Using the term "war" implied mobilization of national resources, focusing public will, and using military force. But the president's declaration of war, backed up by congressional authorization to use military force, meant something greater. In contrast to previous military strikes in response to terrorist attacks, the United States would initiate a *con-*

[I]n the aftermath of the wars in Iraq and Afghanistan, the U.S. military is anxious to avoid large-scale ground deployments to combat terrorists or counter insurgencies.

tinuing campaign against al Qaeda without waiting for further provocations. The U.S. government's aim was no longer deterrence through retaliation, but the destruction of al Qaeda (and its supporter, the Taliban regime in Afghanistan), regardless of the time it would require.

Following the 9/11 attacks, the United States invaded Afghanistan, removed the Taliban government, and dispersed al Qaeda's training camps, but the campaign against al Qaeda soon became conflated with a broader assault on terrorist groups worldwide. Asked in 2002 whether the Global War on Terror would be limited to al Qaeda and the Taliban or would include such groups as Hezbollah, the Deputy Secretary of State responded, "We will bring them all down, one after another." This was hubris.

Many also conflated the Global War on Terror with the continuing efforts to "combat terrorism" by gaining international support for outlawing terrorist tactics. Critics argued that you cannot wage war based on a set of "tactics." In fact, these were two separate, although related, efforts—one to proscribe the use of terrorist tactics, the other to destroy a specific terrorist enterprise.

Public rhetoric further muddied the waters. The Global War on Terror was stretched to include the "axis of evil"—Iraq, Iran, and North Korea—and subsequently to support the invasion of Iraq, which was deliberately portrayed as somehow involved in the 9/11 attacks. Although the invasion of Iraq is now viewed by many as a distraction from the campaign against al Qaeda, the use of military force, including preemptive attacks and targeted killings, is now widely accepted in U.S. policy circles as an appropriate response to terrorism.

And while preferring to concentrate on preparing for conventional wars with near peers, the Pentagon now accepts fighting terrorism as one of its primary responsibilities. Nevertheless, in the aftermath of the wars in Iraq and Afghanistan, the U.S. military is anxious to avoid large-scale ground deployments to combat terrorists or counter insurgencies.

Discussion, Led by General Yishal Beer

Beer led the discussion, which quickly turned into a wide-ranging exchange that encompassed matters of strategy, the role and legality of military operations, the importance of public opinion, the nature of the threat and the tolerance for risk, the influence of casualties on public opinion, and the open-ended nature of today's conflicts. Strategy turned out to be one of the continuing themes of the workshop. The first session ended with an expression of concern that, while terrorists themselves are not yet able to directly threaten democratic states, they might be able to indirectly threaten the survival of democracy.

Beer began by noting that Jenkins' history of American counterterrorism policy showed that there was no strategy; instead, there was an accumulation of ad hoc responses to specific terrorist events—the actions were reactive rather than proactive. That said, there are some good arguments for the United States not seeking a strategy.

Should counterterrorism strategy be based upon what terrorists have done, which can lead to failures of imagination, or what they might do, which allows the United States's and Israel's greatest fears to drive their security?

For one thing, decisive victory does not appear to be achievable in the kind of conflict that the United States and Israel now face—certainly not in the foreseeable future. It therefore would make perfect sense to adopt a strategy of containment, but containment might not assuage public anger caused by terrorist outrages, and it promises no victory. As another participant noted, "defeating" ISIS is impossible. Reducing its capabilities while containing it is feasible. But does that mean accepting the unpleasant prospect of perpetual low-level warfare and, along with it, a continuing terrorist threat? This may be a realistic approach, but it is one that is politically difficult to sell.

This raises the question of what kind of war policymakers and publics are prepared to accept. During the Cold War, containment was considered a preferable alternative to fighting World War III. Today, it might be better to face the daily challenges posed by terrorist adversaries than to escalate the conflict into a costly military contest.

But does the avoidance of large-scale military intervention risk eventual escalation by the other side? One participant cautioned that, although the current levels of terrorist violence seem tolerable, this could rapidly change if terrorists were to acquire weapons of mass destruction. He suggested an enabling, adaptable doctrine rather than a constraining one.

Other views differed. One participant argued that U.S. and Israeli strategy should be balanced, in the sense that it responds to the threat but does not exaggerate it. Another participant cautioned that the United States and Israel should be careful not to construct a counterterrorism strategy based on the worst possible scenario, as it would be unwise to build a U.S. and Israeli entire approach around very unlikely contingencies. This leads to the question, Should counterterrorism strategy be based upon what terrorists *have done*, which can lead to failures of imagination, or what they *might do*, which allows the United States's and Israel's greatest fears to drive their security? Or is the answer something in between, that is, some combination of what the terrorists' intentions indicate and what they are capable of doing? During the Cold War, deterring nuclear war formed the centerpiece of American strategy. To what extent should future scenarios involving terrorist use of weapons of mass destruction drive current strategy?

One participant observed that the religious dimension of the current conflict impeded the formulation of strategy. Noting that most Muslims obviously reject jihadist beliefs, he wondered how a doctrine could be written that accurately and sensitively characterizes the current struggle without referring to its religious dimensions. He concluded that it might be easier to live in a gray area than to codify the true nature of the conflict in which the United States and Israel believe themselves to be. In any case, external actors have only marginal influence in what are major contests within Islam.

Another participant noted that, in current circumstances, Western political leaders might find it difficult to say publicly that their strategy is to allow religious fanatics and their foes or Sunnis and Shias to go on killing each other as long as the violence can be confined to distant deserts and does not wash up on Western shores.

One participant argued that, in an era of weapons of mass destruction, with possibly 200 nuclear bombs in Pakistan, new threats from Iran, and the Sunni challenge, the United States and Israel must have a strategy. The military must be willing and capable to deter and respond to various threats. The challenge

Western democracies are expected to behave according to rules, while the terrorists, by definition, are not expected to do so.

is to develop a strategy that is adaptive. This was easier during the Cold War, when one knew exactly who the enemy was and what had to be done—that is no longer the case. As an example of the need for adaptability, hardly anybody knew of ISIS before its sudden military success in 2014. Therefore, any strategy must be elastic, it must be balanced, and it cannot exaggerate the threat.

An Israeli participant pointed out that there may be one set of principles that govern terrorist strategy, but there is no unifying single strategy. One strategy will not fit all terrorist groups. Each terrorist organization is different—Hamas is not ISIS; ISIS is not al Qaeda; al Qaeda does not seek territory, while ISIS conquers land. There are too many differences.

Beer, like most of the participants, dismissed the Weinberger and Powell Doctrine. It was illusory for militaries to pretend that they were going to commit only when they had a 100-percent chance of success. No country's armed forces have the luxury of choosing their own missions. Even insurance companies—by definition—must take some risks.

An Israeli participant argued that, clearly, some strategy is needed, but the strategy must be innovative. Another Israeli participant argued that the United States had been innovating, most dramatically in the areas of applying sanctions and authorizing the use of military force, suggesting that these two issues are key to thinking about a strategy and that the United States and Israel should reflect on the inherent differences between the two strategies and the reasons for their adoption. Other participants suggested that the United States and Israel overestimate the extent to which there are military solutions to these problems.

Legal issues arose several times in the first session, with some participants arguing that, in some cases, the United States has acted outside the acceptable norms of international law. In 1986, two American soldiers were killed in a nightclub, and, in retaliation, American warplanes bombed Libya. Beer said that this bombing was unlawful. After the attempted assassination of former president George H. W. Bush in 1993, American planes bombed Baghdad, which was also illegal, according to Beer.

Other participants suggested that the Bush Doctrine of *preemption* was really just *prevention*, which is illegal under international law. One participant countered that the United States could find ways to legally justify almost any military action and was currently engaged in "calisthenics" to fit its military operations into current legal structures. It is not merely a matter of law, one participant pointed out, but a broader issue of legitimacy—that is, was the action perceived as "just"? World War II was, but the invasion of Iraq and the Gaza operation were not. International acceptance is crucial, but that means that there is an asymmetry in respect for rules and values, which is more important than the disparity in weapons and tactics, countered another participant. Western democracies are expected to behave according to rules, while the terrorists, by definition, are not expected to do so.

Reprisals are illegal under international law, but different regimes can apply. A number of the legal scholars at the workshop allowed that certain tacit policies might be interpreted as an "acceptable compromise" under the circumstances. Participants returned to legal arguments in Session Five, while the tension between public policy and its discretionary execution remained a recurring theme in the conversations.

Participants emphasized that public opinion plays an enormous role in contemporary conflict, particularly in Western democracies. An American participant noted that public opinion is even more

> *Expectations of absolute security and demands for vigorous action put enormous pressure on state institutions that cannot afford a single terrorist attack and must strike back hard if one occurs.*

important now than it was during the Vietnam War, and, in general, the public is much better informed now than it was at that time. North Vietnam's war strategy focused on American public opinion. The North Vietnamese tried to undermine the political will of the United States by making America look like the villain. Communications play an even more important role in ISIS's strategy, but one cannot always discern the group's intentions. Is its use of social media to publicize its atrocities intended to dissuade the United States from becoming directly involved in the conflict, or is it intended to provoke American military intervention?

Another American participant cautioned against trying to understand public opinion's role in military strategy, as such dynamics are notoriously hard to understand and predict. He noted that public opinion can make sudden swings. As was recently illustrated, two publicized beheadings can change American attitudes and policy.

Casualties have significant influence on public opinion. Eric Larson discussed his past research on the role of casualties and public opinion, identifying the factors that have been shown to be consistently important. These include whether the public viewed the military mission as something worth doing. Was it "winnable"? That is, could it be done? Were the nation's leaders fractured or in agreement? Were the United States and Israel making progress? Were objectives changing? Were casualty rates increasing or declining? How the public sees these issues significantly impacts its willingness to accept casualties. Understanding this enables us to predict the path of public opinion, but the results are not easily exploited.

Casualties are one dimension, but also important is the increasing role of the public's appetite for action and tolerance for risk. Some participants noted parallels with historical campaigns, many of which had been brought to an end, asking why the United States's and Israel's present conflicts seemed so perpetual and unending. They suggested that part of the difference was that other societies had acquiesced to certain "acceptable" levels of risk, which the United States and Israel have been unwilling to countenance.

Expectations of absolute security and demands for vigorous action put enormous pressure on state institutions that cannot afford a single terrorist attack and must strike back hard if one occurs; these expectations also put enormous pressure on political leaders, strongly influencing what the leaders feel they must deliver to remain in power.

The session closed with Ayalon worrying aloud about the fact that as a consequence of the threats it faces, Israel itself is changing. How long can Israel remain a democracy? He also expressed his concern that the changes were affecting how Israel is being portrayed internationally.

CHAPTER FOUR

Session Two: Distinctions in Asymmetric Warfare

Opening Remarks by Amichay Ayalon[1]

Before one can develop a strategy, one must understand the framework. Presently, the United States and Israel know the details of the threat and generally understand it, but the nations have no framework or doctrine. Ayalon presented his research on developing a general theoretical framework for understanding and explaining the phenomenon of terrorism. In this work, he sought to develop a rubric by which to distinguish between terrorist groups and better understand their structures and behaviors. He also suggested its application to the study of violent nonstate actors in order to predict and influence their behavior, evaluate them comparatively, and "facilitate their coherent and systematic study."

Ayalon's study attempts to reshape the way the United States and Israel think about terrorism in general and violent nonstate actors in particular. In order to accomplish this, the nations must replace the contemporary preference for complexity with a return to the basic categories of structure and behavior.

The structural explanation, built on Carl von Clausewitz's trinities as outlined in *On War* (Clausewitz, 1976), emphasizes the continuing existence of the three basic building blocks shown in Figure 4.1.

Reaffirming the essentially unchanging nature of war, Ayalon carefully traces the evolution of government, army, and people from the pre-Napoleonic ideal types to today's amorphous entities. This change, beyond the blurring of the traditional boundaries and the generation of overlapping areas, generates a partial or general decoupling between structure and function, as shown in Figure 4.2.

According to Ayalon, the functions of government, army, and people are very much observable in each and every one of today's nonstate actors. The difference between these manifests itself in the structures generating three archetypal violent nonstate actors: (1) monadic, in which all three of the abovementioned functions are contained in one amorphous structure; (2) dyadic, in which the government's structures and functions are more or less congruent, while the functions of people and army cohabit in a second amorphous structure;

> *Ayalon's study attempts to reshape the way the United States and Israel think about terrorism in general and violent nonstate actors in particular.*

---

[1] This presentation is based on an unpublished paper by Ayalon, Robert C. Castel, and Elad Popovich.

Figure 4.1
Basic Building Blocks of Terrorism

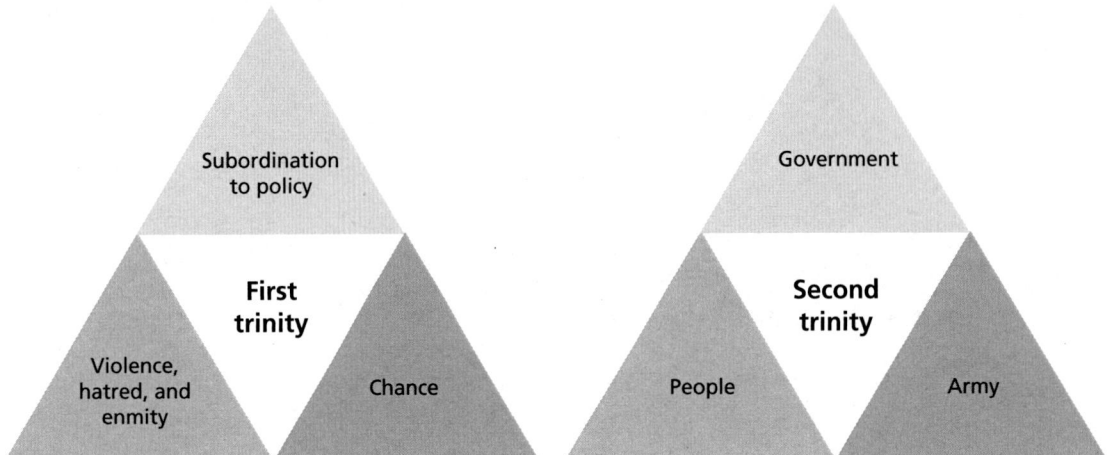

SOURCE: Unpublished paper by Ayalon, Castel, and Popovich.
RAND *CF334-4.1*

Figure 4.2
The Evolution of Government

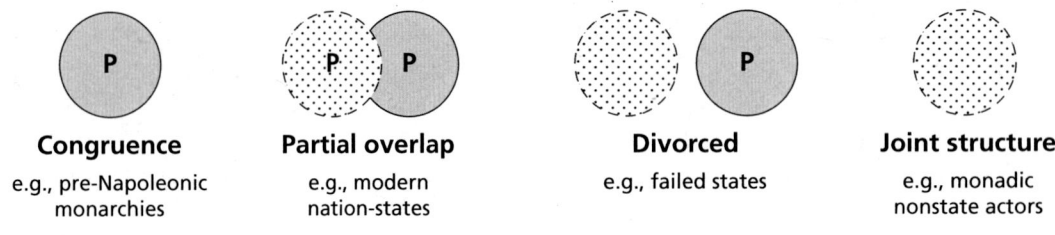

SOURCE: Unpublished paper by Ayalon, Castel, and Popovich.
RAND *CF334-4.2*

and (3) triadic, where the three functions are mirrored by three structures, albeit in a much less clear-cut way than the Clausewitzean ideal type.

In Ayalon's view, these three structures generate (1) three distinct behavioral patterns and (2) three distinct stages in the violent nonstate actor's evolution toward statehood.

In order to offer a better explanation of violent nonstate actors' behavior, Ayalon borrows two powerful dichotomies. The first is Robert K. Merton's analysis of conformity versus nonconformity, while the second is Max Weber's two types of ethics, that of responsibility and that of ultimate ends. Combining them into a classical two-dimensional typology generates a powerful tool for analyzing violent nonstate actors' behavior, as well as its evolution (Figure 4.3).

- Quadrant 1: Violent nonstate actors in this quadrant are characterized by nonconformism in their goals, as manifested in their ambition to change the international system or at least disconnect themselves from it. Their political conduct is characterized by responsibility, as manifested mainly in the care of the population for which they bear responsibility.

Figure 4.3
Typology of Violent Nonstate Actors' Behavior

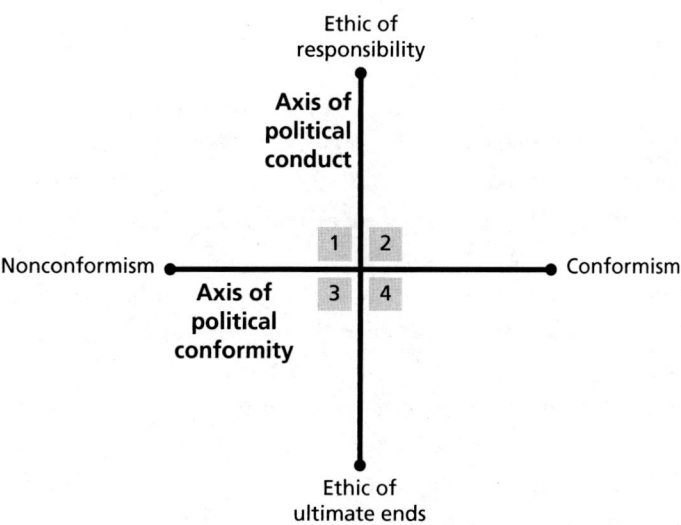

SOURCE: Unpublished paper by Ayalon, Castel, and Popovich.
RAND *CF334-4.3*

- Quadrant 2: Violent nonstate actors in this quadrant live in a world of conflicting values, and decisions regarding the hierarchy of these values are made on a case-by-case basis. These violent nonstate actors are characterized by conformism in their goals, as manifested in their motivation to join the international system rather than transform it. Their political conduct is characterized by responsibility, as manifested in their use of means that are at least partly congruous with the means deemed acceptable in the international system. Another characteristic is the emergence of gaps between their proclaimed ideology on the one hand and their observable behavior on the other.

- Quadrant 3: Violent nonstate actors in this quadrant live in the same reality of conflicting values, but their decisions regarding the hierarchy of these values is preordained, sometimes by a divine authority or its secular equivalent. This clarity of purpose makes the dialogue between strategy and diplomacy largely superfluous. Violent nonstate actors in this quadrant are characterized by nonconformism in their goals, as manifested in their ambition to destroy the international system and remake it in their own image. Their political conduct is characterized by an adherence to ultimate ends and a lack of responsibility toward any kind of population.

- Quadrant 4: Violent nonstate actors in this quadrant are characterized by conformism in their goals, as manifested in a lack of ambition to change the international system. However, their political conduct is characterized by adherence to ultimate ends and a lack of responsibility toward their population.

In Ayalon's model, the combination of the two typologies suggests that certain organizational structures tend to generate certain patterns of behavior. For instance, monadic and dyadic violent nonstate actors tend to cluster around quadrant 3, and their evolution toward a triadic form is accompanied by a migration toward quadrant 2. Putting it differently, a small, brutal terrorist group, by its evolution into a larger guerrilla organization, will most likely turn into a more responsible entity along the way.

Ayalon used the example of the Muslim Brotherhood to illustrate how a violent nonstate actor can move among quadrants. The Muslim Brotherhood started in Egypt in the 1920s and was more interested in *Da'wah*—education and charity—than in expelling the British militarily from Egypt, so Ayalon would initially place the group in quadrant 2. As the organization adopted violent terrorism under the guise of Pan-Arab ideology, it entered quadrant 3. With the victory of Mohammed Morsi in Egypt in 2012, the Muslim Brotherhood moved firmly to quadrant 1. Now, with the military takeover of Egypt in 2013, the Muslim Brotherhood is at a critical juncture between quadrants 1 and 3.

Ayalon also argued for the importance of assessing whether groups viewed their responsibilities as being global or simply to their own people and the implications of this distinction for understanding their actions. He suggested that organizations in quadrant 2 tended to be more pragmatic. Ayalon also distinguished between Islamist groups that take long-term, gradualist strategies of education and charity and those such as ISIS, which move with much more immediacy. He cited the letters from al Qaeda leader Ayman al-Zawahiri to the leader of al Qaeda's affiliate in Iraq, Musab al-Zarqawi, regarding the latter's ruthless strategy toward other Muslims and Zawahiri's implicit acknowledgment of reliance on people for legitimacy and power.

This same diagram could be used as a framework for understanding an organization such as Hamas and the interplay within its organizational structure. Ayalon cautioned that observers seeking to understand organizations should look not only to their covenants, as these do not always describe their behavior well, but also to their actions, arguing that Hamas had transitioned from being a quadrant 3 to a quadrant 2 organization, as it had become increasingly conformist to international goals and responsible to its constituency.

Ayalon suggested that these observations called for redefinition of the concept of deterrence, as well as more reflection on traditional understandings of warning, decisiveness, victory, strategy, diplomacy, and international humanitarian law.

The point of this analysis, Ayalon emphasized, was to better enable democracies to defeat violent nonstate actors without betraying the values that the democracies are trying to preserve.

Discussion, Led by Andrew Liepman

Liepman agreed with Ayalon's main argument about the importance of moving away from the tendency to view all terrorist groups as the same. They are not, and it is important to make distinctions, although this thinking runs against an American tendency to see all issues in terms of binary distinctions—in this case, good versus evil.

> Unless the United States is committed to unilaterally destroying all of the terrorist groups in the world, it must in each case ask, What is the nature of the threat to the United States?

Although, as a matter of principle, the United States opposes terrorism and has a long record of enlisting international cooperation to combat the use of terrorist tactics by anyone, that cannot be the basis for its strategy in dealing with specific terrorist organizations. Some of the most violent extremist groups have little immediate relevance to the United States. Unless the United States is committed to unilaterally destroying all of the terrorist groups in the world, it must, in each case, ask, What is the nature of the threat to the United States? How much do the United States and Israel care about a specific group? And how much should they care?

[W]hile strategy has to be important, caution is in order. The history of modern terrorism is actually a chronology of discrete events.

As part of an attempt to answer these questions, American intelligence analysts tried to plot terrorist organizations on a chart with one axis representing the group's capabilities and record of violence and the other axis charting the degree to which the group focused on the United States. For example, while Boko Haram poses a threat to Nigeria and the surrounding areas, it is not presently a direct threat to the United States.

Other participants later pointed out that, while not all terrorist groups pose a direct threat to the United States, they threaten U.S. allies and regional stability, which makes them a concern. Clearly, the application of these theoretical models to reality is a major challenge.

Liepman added that the enemy is not static, but rather is constantly evolving and may change its strategy to confront new circumstances. As an example, al Qaeda in the Islamic Maghreb is not a direct threat to America, but the group could become one if the United States were to attack it.

Liepman ended his comments with another question: Do the United States and Israel have the right metrics to measure progress against these groups? Removing their leadership is important, and it is something that can be quantified, but clearly it is not the only or necessarily the most important measure. Other participants agreed that analysts tend to measure what they can rather than what is truly important.

This initial exchange set off another lively discussion, much of it revolving around the question of the extent to which the United States and Israel should strive for a grand strategy. Most agreed that, while strategy has to be important, caution is in order. The history of modern terrorism is actually a chronology of discrete events. The perpetrators are a diverse bunch. The grand strategies that arose out of these events may have served political, diplomatic, and military purposes at the moment, but ultimately they did not prove conducive to clear thinking and coherent planning, and sometimes they even got in the way. Often, strategy comprised little more than desiderata—a catalog of lofty goals with no instructions on how they could be achieved.

Analysts do not invent terrorist tactics—the attacks are real—but, to a large extent, the phenomenon of terrorism is an analyst's construct. We make boxes to categorize terrorist events and terrorist groups, but these are our inventions to serve a nation's goals, for example, enlisting international cooperation against terrorist tactics and groups. Theory can too easily become the prism through which one views each threat. Driving toward a global grand strategy may get in the way of tactical flexibility.

Some participants cautioned that a solid strategy, while remaining ultimately rational, cannot escape the emotional component. The public may not always exhibit stoicism or patience; it often demands a response to terrorist provocation (such as the beheadings of James Foley and Steven Sotloff). Domestic politics often determine strategy. And sometimes governments themselves want to erase distinctions for political ends. For example, saying that ISIS is Hamas and Hamas is ISIS—in other words, that there is no difference in their tactics—helps mobilize allies.

One participant suggested that Ayalon's model is useful for distinguishing between terrorists and insurgents, particularly in examining whether the use of terror is incidental or central to their raison d'être. It is desirable that these groups become more responsible but not more capable. That led to a discussion of the group's goals.

If it is difficult (and perhaps even unwise) to formulate a grand strategy to defeat terrorism, certainly the United States and Israel need to be clearer about how they deal with specific groups—for example, al Qaeda or ISIS. But if strategy depends on correctly categorizing a group, how should the United States and Israel categorize ISIS? Participants debated whether the group's goal is territory or a theological ideal. ISIS is a sort of hybrid. It routinely employs extreme violence, uses terrorist tactics, and carries out atrocities to further its ideological goals, but, at the same time, it seeks to capture territory, govern populations, collect taxes, run courts, support social works, provide education, and build a permanent state.

But does that make ISIS an anomaly? History is filled with brutal armed conquests involving tactics that would today be labeled as terrorist, followed by armed occupation and nation-building. Or is this an example of where the terrorism framework confines our thinking about strategy?

In the discussion, some participants noted that holding territory imposes responsibility, or moving to quadrant 3—political conformity—on Ayalon's chart. Does that mean that ISIS eventually will be tamed by the requirements of governance? Some participants thought so. One participant pointed out that, when the homeless become landlords, their outlook changes—Hezbollah provides an example. Others disagreed that Hezbollah had been tamed.

Another key question with regard to ISIS was whether it is a group whose behaviors and actions could be shaped in such a way as to move its leaders toward a political dialogue or whether it remains ultimately "irretrievable." Some participants stressed that a key dynamic is the direction of the group's evolution and whether it is progressing toward a position more amenable to negotiation or less. They emphasized that this evolution is not independent but rather is affected by U.S. and Israeli responses.

Others noted that the manner in which ISIS takes control leads to persistent questions about its legitimacy. These issues may make ISIS a serious anomaly in the proposed characterization schematic. Values must be taken into account, one Israeli participant noted—the Nazis wanted peace after conquering most of Europe. Another noted that ISIS rejects international law. Because it is a terrorist group, do the United States and Israel hold it to a lower standard?

A RAND participant made the point that groups can become more responsible as they move to the right of Ayalon's quadrants, but that they may also become more capable as they do so. Some may become more amenable to negotiations, while others evolve into greater threats. Ultimate goals are important. Hamas has limited territorial aspirations, while ISIS has the entire Middle East in its sights.

Larson brought up the different competing groups within a move-

> *If it is difficult (and perhaps even unwise) to formulate a grand strategy to defeat terrorism, certainly the United States and Israel need to be clearer on how they deal with specific groups.*

ment, some of which may be redeemable, while others are not. One strategy may be to peel off the "redeemable" elements from the ultras, which will never accept compromise.

What Israelis miss is that others in the world will not perceive things the same way they will, and that can cause damage to the country's reputation and its economy.

Jenkins asked how the Sunnis in ISIS-controlled territories should be characterized: Are they hostages of ISIS or supporters? The answer might depend on how the United States and Israel choose to categorize them, and that may depend on tactical expediency. If their eventual support is necessary, the United States and Israel may choose to treat them as victims of occupation rather than active collaborators. Jenkins provided the example of Austrians during World War II.

Ayalon had the final word and offered several observations. Mistakes can arise from strategies based on simplistic or ideologically driven analysis. As an example, a number of American leaders were convinced that, just as the Americans had rebuilt Europe and Japan following World War II, America could rebuild the Middle East in its own image by destroying tyranny and replacing it with democracy.

In past wars, the fight was conducted on the battlefield, following which the victors and the vanquished sorted out the terms of peace. In contrast, the current conflict could last another 40 years, and it will be sorted out not on the battlefield but rather through the lens of the media. Victory will be in the eyes of the spectators, and they will affect government policy. Success on the battlefield can lead to strategic failure.

It is a matter of perceptions. Just as Americans may be victims of their illusions about the easy exportability of democracy, Israelis are emotional about terrorism. This can propel them to take actions that are counterproductive. Ayalon used the example of stopping the 2010 Turkish flotilla that was traveling to Gaza. Stopping the flotilla was a tactical victory, but it ended as a strategic defeat. What Israelis miss is that others in the world will not perceive things the same way they will, and that can cause damage to the country's reputation and its economy. Ayalon said that the United States and Israel can kill the enemy, but, if U.S. and Israeli values have been trammeled and their economies are in ruins as a consequence of boycotts and sanctions, the countries lose.

Ayalon said that many of the questions raised would be the subjects of our future collaboration. Contemporary conflict must take into account four different (but overlapping) campaigns, or fronts: military operations, diplomacy, media (or what we might call "image-fare"), and battles over how what law applies or how it should be applied (which we might call "lawfare").

CHAPTER FIVE

Session Three: RAND Research on Counterinsurgency and Counterterrorism

Opening Remarks by Colin Clarke, Patrick Johnston, Rick Brennan, Eric Larson, and David Johnson

Session Three offered an opportunity for RAND researchers to briefly review recent, ongoing, and prospective RAND research on counterinsurgency and counterterrorism. Clarke led off the session with a discussion of research on sources of success in counterinsurgency, published in a 2013 monograph (Paul et al., 2013; Paul, Clarke, and Grill, 2010). This research examined all insurgencies begun and ended between the end of World War II and 2010 and quantitatively tested the performance of more than 20 separate counterinsurgency approaches, which were taken from a survey of the existing literature.

The monograph was written with the intention of providing an analytical framework for a host government facing an insurgency. Factors that assisted or hindered the counterinsurgency forces in each case study and each phase were examined by identifying common factors. The research looked at cases in South America, Africa, the Balkans, Central Asia, and the Far East. Researchers also looked at operations to contain the fight against South African apartheid, as well as anticolonial rebellions.

Clarke said that the research had identified three essential ingredients for successful counterinsurgency operations:

- flexibility and adaptability on the part of the counterinsurgency force
- impeding the flow of arms and other forms of tangible support to the insurgents
- commitment and motivation of the host-nation government.

RAND researchers developed a "scorecard" of good and bad factors in order to assess each case examined. In the example of the post-Soviet Afghanistan insurgency, the counterinsurgency strategy failed to address the "good" factors identified by the RAND team and scored high on the "bad" factors. The research also looked at how external actors affect success in counterinsurgency operations, finding that combatants cannot want victory more than the local population.

Johnston followed Clarke with a brief overview of a RAND research project focusing on ISIS finances and organization (Bahney et al., 2010). Here, researchers examined al Qaeda in Anbar Province during 2005 and 2006, when it was considered to be at the peak of its control. The study relied on captured documents, including al Qaeda PowerPoint presentations and ledgers, for its analysis and conclusions. Al Qaeda's finances were tied to its organizational structure, as shown in Figure 5.1.

Figure 5.1
Anbar Province Revenues: June 2005–May 2006

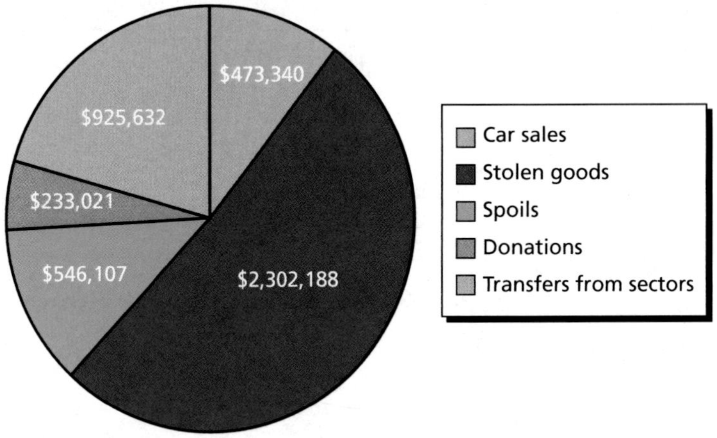

SOURCES: Brother 'Imad, 2007a, 2007b; and two other 2007 Harmony
Database documents: NMEC-2007-633541 and NMEC-2007-633893.
RAND *CF334-5.1*

Johnston showed a pie chart indicating the sources of al Qaeda's revenue in Anbar Province from June 2005 to May 2006. The chart comprised five segments: car sales, stolen goods, spoils, donations, and transfers from sectors. The segments are those selected by al Qaeda's own bureaucrats, not the RAND researchers. More than half of the annual total revenue came from stolen goods.

Using a bar chart (Figure 5.2), Johnston then showed that revenue fluctuated widely from month to month. Researchers found that most of the money came from self-financing activities, not from other countries, such as Qatar. (In the discussion that followed, some of the Israeli participants expressed surprise that the Gulf States were not funding ISIS.) The captured records also show meticulous record-keeping.

In contrast, documents captured in Mosul show that revenue was generated mostly from the sale of oil and extortion (al Qaeda called this "contracting work"), not petty crime.[1]

Asked about the banking system in today's Islamic State, Johnston responded that the Islamic State operates a cash economy, with ISIS spending the money as it comes in—not much cash is kept on hand. Coalition bombing has cut ISIS's revenue from the sale of oil on the black market by half.

The third RAND presentation was by Larson and focused on relevant research on counterinsurgency and counterterrorism. Larson noted several areas in which RAND has conducted research and offered highlights:

- counterinsurgency and irregular warfare
- al Qaeda's strategy, narrative, and discourse
- influence operations
- social science theory of intelligence tradecraft
- opinion leadership, the media, and public opinion during U.S. military operations.

---

[1] Johnston's findings were summarized in Johnston and Bahney, 2014, and in his testimony before the House Financial Services Committee on November 13, 2014 (Johnston, 2014).

Figure 5.2
Financial Foundations of the Islamic State: Anbar Province, 2005–2006

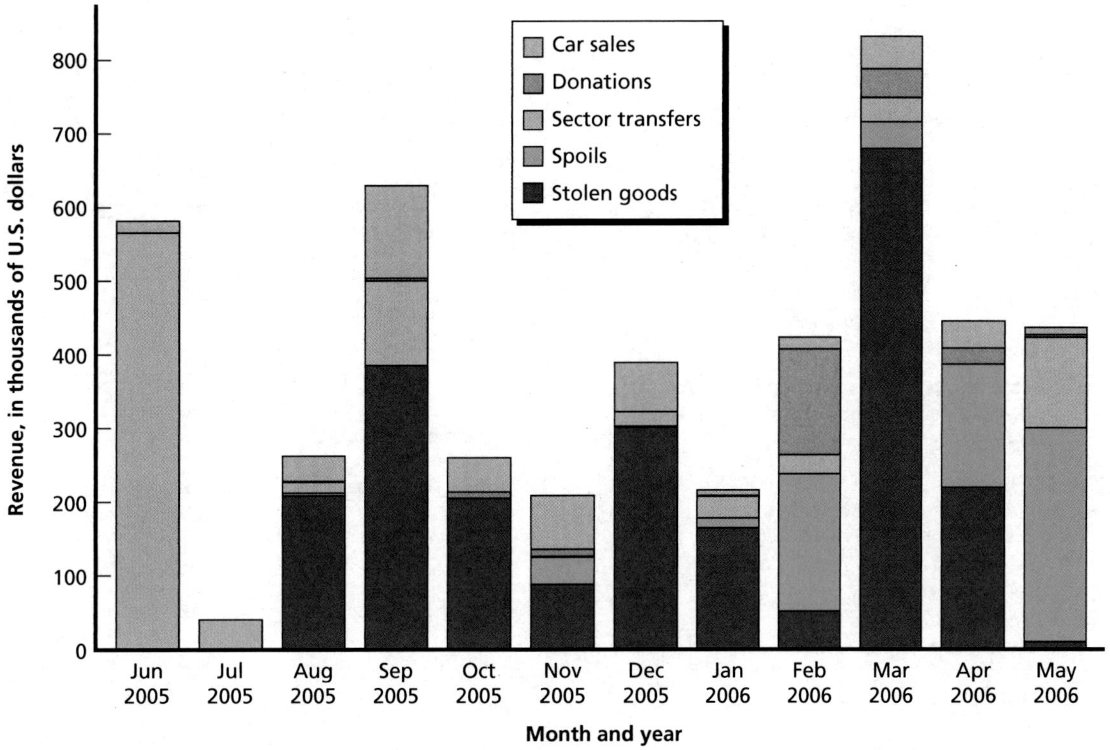

SOURCES: Brother 'Imad, 2007a, 2007b; and two other 2007 Harmony Database documents: NMEC-2007-633541 and NMEC-2007-633893.

RAND *CF334-5.2*

Larson started with Leites and Wolf on counterinsurgency strategy (Leites and Wolf, 1970). In *Rebellion and Authority*, Leites and Wolf argued that it was necessary to focus on "supply" issues such as eroding an adversary's capabilities and "demand" issues such as winning "hearts and minds." One way to think about "hearts" is that the population will naturally support the more human party, while "minds" relates to which side the population thinks is more likely to win, since few want to be caught "on the wrong side of history." Some general observations are presented in Figure 5.3.

Some asymmetries favor the insurgents: will and motivation, external support, suitable strategies and tactics, suitable forces, time discount rates, and knowledge of the human battlespace. In counterinsurgency operations, democracies are almost always at a disadvantage, and democracies lack patience to fight long conflicts.

Talking about "the terrorists" is meaningless—it is important to differentiate among terrorist groups. There are differences between al Qaeda central and its subsidiaries, which themselves differ from Iraq to Pakistan to Afghanistan to Yemen, and differences between the al Qaeda trend and other groups, such as Hezbollah or Hamas. It is important to understand distinct trends and the composition of forces and their relationships, as well as the strategy, narratives, and discourse of each group, in order to analyze strengths, weaknesses, potential wedge issues, propaganda aims, and efficacy.

Larson then focused on the role of the media and public opinion. He noted that the media have historically been strong supporters of various U.S. military operations. Media reporting

Figure 5.3
Observations on Counterinsurgency and Terrorism

<table>
<tr><td>

On Counterinsurgency

- The essentially political nature of counter-insurgency, some terrorism
  - Insurgency/counterinsurgency intrinsically political
  - Political groups with militant arms
  - Terrorist groups with political fronts
- Leites and Wolf (1970) on counterinsurgency strategy
  - "supply": erode adversary capabilities
  - "demand": hearts and minds (HAM)
  - Both simultaneously
- Thinking about "hearts and minds"
  - "Hearts": Who will be more humane?
  - "Minds": Who will win?
- Asymmetries that favor the insurgent
  - Will, motivation (Mack)
  - External support (Byman, Record)
  - Suitable strategies, tactics (Arreguin-Toft)
  - Suitable forces (Lyall)
  - Time discount rates
  - Knowledge of the human battlespace

</td><td>

On Terrorism

- Unpacking "terrorism" into distinct trends, groups, understanding their relationships (e.g., Salafi-jihadi versus Muslim Brotherhood)
- Understanding distinct strategies, narratives, and discourse of each group as a lens into group strengths, weaknesses, wedge issues, propaganda aims, and efficacy

On Media and Public Opinion

- Media as "government's little helper"
- Conditional nature of media effects
- "Misfortunes of War" (civilian deaths)
  - Media effects: Is it the fact of civilian deaths or the media's reporting of these deaths that matters most?
  - Attitudes: Belief U.S. making every effort to avoid civilian casualties leads to higher support for military operations
  - Transparency of efforts to ensure targeting complies with laws of war

</td></tr>
</table>

SOURCE: Larson.
RAND *CF334-5.3*

of civilian deaths can have an impact on U.S. public opinion, but the impact is muted to the extent that the public understands that the military is making every effort to avoid civilian deaths.

Larson then presented a construct for analyzing terrorist or insurgent organizations and assessing their prospects for building support (Figure 5.4). It includes five levels: strategic objectives and decisionmaking, ideology, frames and framing processes, resource mobilization, and political opportunities and constraints. With this construct as a framework, al Qaeda's strategic objective can be seen as trying to restore the Caliphate. It draws from salafi-jihadi ideology, theology, and jurisprudence for its apologetics. Al Qaeda frames its basic appeal for support in terms of resisting the oppression of Muslims and an alleged war on Islam. Al Qaeda's resources are mobilized through organizational structures, including operations, finance, logistics, propaganda, recruiting, and other key functions. Al Qaeda is attempting to survive in the face of a hostile political-military environment but also to exploit opportunities to grow the movement.

Larson then turned to a review of scholarship on the role of messaging and "media effects." Sweeping generalizations about the role of the media in shaping attitudes (e.g., the so-called "CNN effect") have generally been found lacking. Such effects are highly conditional and depend on factors that influence whether messages reach the target audience, such as media "filters," and individual awareness and individual "filters" that regulate whether messages are accepted and whether they will affect beliefs, attitudes, preferences, and behaviors (Figure 5.5). Even in cases in which such effects can be detected, the effects are typically modest.

Figure 5.4
Framework for Analyzing a Terrorist Group's Prospects for Building
Support, Using al Qaeda as an Example

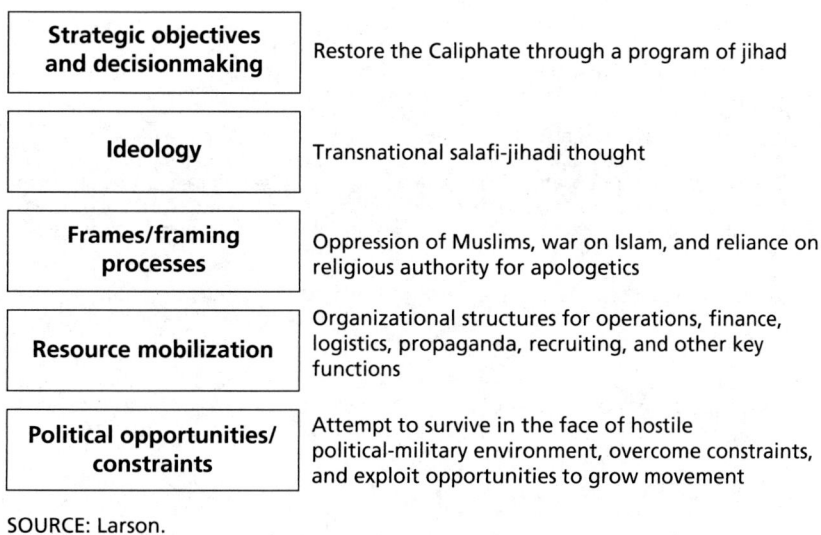

SOURCE: Larson.
RAND *CF334-5.4*

Figure 5.5
Effects of Messaging and Media on Public Attitudes

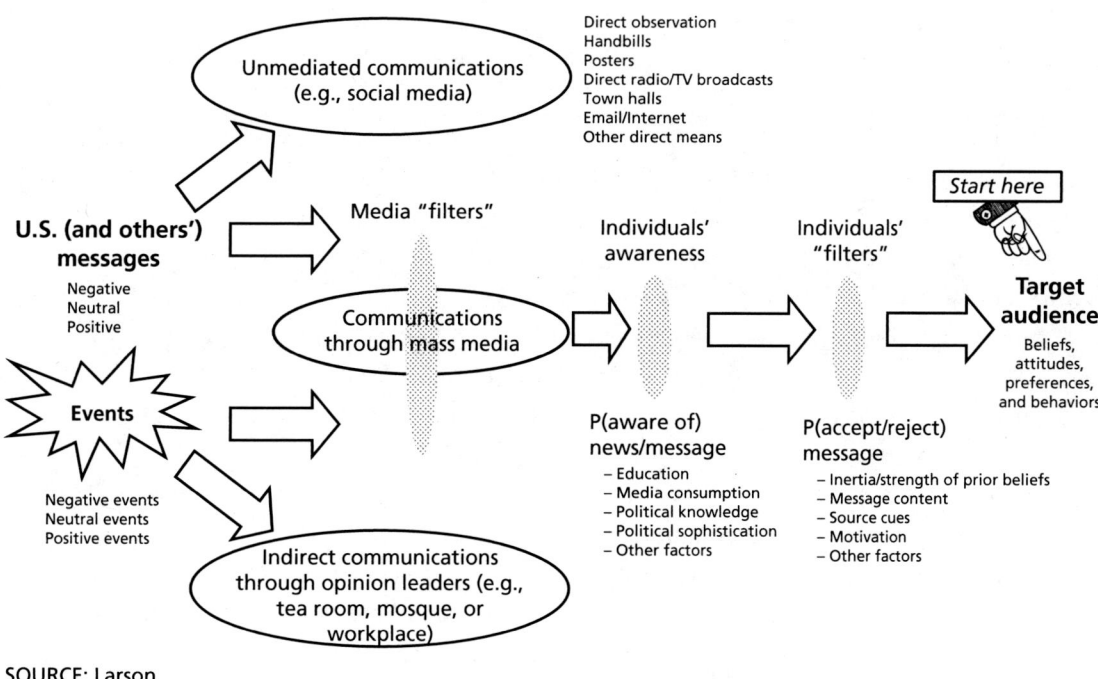

SOURCE: Larson.
RAND *CF334-5.5*

Some Possibly Surprising Findings on "Oppositional Media" and "Media Effects"

Scholars such as Hallin (1989) and Zaller (1999) have found that the U.S. mass media have more often served as "government's little helper" than its adversary during crisis and war.

In *The Nature and Origins of Mass Opinion* (1992), John R. Zaller found political sophistication, partisanship, and whether a policy area involved one- or two-sided messaging to be strong predictors of political and policy opinions but was unable to say with any certainty whether mass media reporting or personal networks were more important influences.

"This study investigates the differential effects of exposure and attention to news and of interpersonal communication on change in public opinion under the condition of one-sided or two-sided information flows. . . . The results confirmed our hypotheses and showed media effects for less politically sophisticated individuals under the condition of a one-sided message flow and effects of interpersonal communication for politically sophisticated individuals. Media had no effect under the condition of a two-sided message flow." (de Vreese and Boomgaarden, 2006, p. 19)

"There is a huge difference between public perceptions of the power of media in elections and academic evidence of its influence . . . . Public beliefs in omnipotent media contribute to wasted time and money; ultimately, they undermine the legitimacy of election outcomes." (Mutz, 2012, p. 83)

"This article examines if the emergence of more partisan media has contributed to political polarization and led Americans to support more partisan policies and candidates. . . . [E]vidence for a causal link between more partisan messages and changing attitudes or behaviors is mixed at best." (Prior, 2013, p. 101)

SOURCES: Hallin, 1989; Zaller, 1992, 1999; Vreese and Boomgaarden, 2006; Mutz, 2012; Prior, 2013.

To understand media effects, one can map media-influence networks:

- key political factions and groups and other trends
- their media outlets
- messages and message sources
- audience survey data.

The fourth RAND presentation, by Johnson, examined the range of military options based on the future security environment. Potential adversaries can be divided into three broad categories: nonstate irregular adversaries; state-sponsored hybrid adversaries; and traditional, or regular, state adversaries. Future U.S. capabilities should be linked to potential adversary capabilities across the full range of military operations.

Johnson pointed out that, although irregular forces tend not to be well trained, are lightly armed, and operate with little discipline under informal command and control procedures with limited communication capabilities (cell phones and runners), they still can be very effective. Irregular adversaries can, however, make a rapid transition to a "hybrid" adversary when a state provides them advanced standoff weapon capabilities. This is what happened when the United States supplied advanced weapons to the Afghan Mujahideen (Stinger man-portable air-defense systems) that were effective in shooting down Soviet aircraft. This standoff capability took away the Soviet air mobile capability and created conditions that made it impossible to operate beyond roads. Additionally, the ability of the Soviets to stop the flow of weapons from Pakistan was daunting. By some estimates, it would have taken an additional 600,000 Soviet troops to block Mujahideen supply lines.

At the other end of the continuum are "high-end" state adversaries. Some of the examples Johnson provided were the Soviet Union in Afghanistan in the 1970s and 1980s, Russia in Chechnya in the 1990s, Israel in Lebanon in 2006, and the United States in Afghanistan and

Iraq. In contrast to nonstate adversaries, these state adversaries tend to be hierarchical, with sophisticated weapons and generally centralized control.

State-sponsored hybrid adversaries are not ordinary insurgents. Examples would include Afghanistan's Mujahideen, active during the 1980s; Chechen rebels; and Hezbollah in Lebanon. Organizationally, the hybrid entity is moderately trained and disciplined. The weapons are light, but the forces have standoff capabilities (for example, antitank guided missiles, man-portable air-defense systems, and longer-range rockets). Command and control is typically decentralized.

The United States has not confronted hybrid adversaries since the Vietnam War. Conflict with state-sponsored hybrid entities presents high-intensity combat challenges that require joint, combined-arms fire and maneuverability. Johnson characterized ISIS as a model similar to the Chechen rebels in 2006.

Discussion

The presentations prompted numerous comments on the various issues raised. Jenkins began the discussion, noting that local knowledge is critical to counterinsurgency operations. He emphasized the importance of "retail politics" in any counterinsurgency campaign and questioned whether the United States and Israel should be looking for a grand strategy or local strategies tailored to each situation. Ayalon agreed that everything is local. (This turned out to be a continuing theme in the discussions.)

Several examples were added to Johnson's catalog of adversary types. Hezbollah's attack on the Israeli Corvette-class navy ship in 2006 was viewed as a demonstration of hybrid power. Recent Russian actions in Ukraine may be another example of a hybrid adversary.

There was some consensus that it has become more difficult to fight insurgencies in the past 20 years. Things have changed. For one thing, the media and especially the arrival of the Internet have changed in the past 25 years. Live coverage has been extended onto the battlefield, something first seen during the Vietnam War. The hunger of competitive 24/7 news networks for revelations and controversy has complicated communications and the ability to control narratives. The Internet has increased the ability of adversaries to directly engage with their perceived constituents and broader audiences. Ayalon pointed out that although analysts tend to understand the media the old way, it is clearly different today, in part with the development of technology like the now-ubiquitous cell-phone cameras, as well as the expansion of social media. But, as Larson pointed out, the effects of the media on public opinion are often only modest.

Brennan added a cautionary note regarding the ability of government to shape public opinion. He pointed out that the United States tried to shape public opinion in Iraq, but it proved to be not very effective—the Iraqis disliked the Americans only slightly less than they

> [A]lthough analysts tend to understand the media the old way, it is clearly different today, in part with the development of technology like the now-ubiquitous cell-phone cameras, as well as the expansion of social media.

disliked the Iranians. Local politics determine public attitudes. It is the behavior of the local government that will shape public opinion.

Jenkins added the example of Vietnam. The South Vietnamese government, which visibly depended on the Americans, was hugely corrupt at the national and local levels. The American style of warfare destroyed the countryside, while Americans—foreigners—were killing thousands of Vietnamese. The average Vietnamese peasant on the ground was not necessarily pro-Vietcong but understandably did not see the Americans as saviors. Michael Spirtas added that the rapid advance of ISIS across northern Iraq in 2014 should not have been surprising given the level of Sunni discontent in the country. Spirtas said, "We knew who Maliki was after six months. Why were we surprised in 2014?"

Ayalon noted that Israel repeats these mistakes. The Israelis imagined that they were the good guys and compelled to respond militarily in Gaza only when there were no other alternatives. From this perspective, the residents of Gaza would understand that they suffered because of the actions of their Hamas leaders. With the cease-fire, however, the population of Gaza was found to be supporting Hamas even more, despite the devastation. After 49 days of war, Hamas's approval ratings went from 22–23 percent to more than 60 percent. The approval rating for Palestinian leader Abbas fell to 31 percent.

The discussion provided a segue to the topic of "imagefare" in Session Four.

CHAPTER SIX

Session Four: From Warfare to Imagefare

Opening Remarks by Moran Yarchi[1]

Moran Yarchi began her presentation with a quote from President Barack Obama, speaking at the U.S. Naval Academy on May 24, 2013:

> In our digital age, a single image from the battlefield of troops falling short of their standards can go viral and endanger our forces and undermine our efforts to achieve security and peace.

Yachi said that the goals of her study were twofold:

- to analyze the level of asymmetry between counterparts in a conflict
- to examine foreign media coverage of the conflict (see Figure 6.1).

Figure 6.1
Foreign Media Coverage of a Conflict

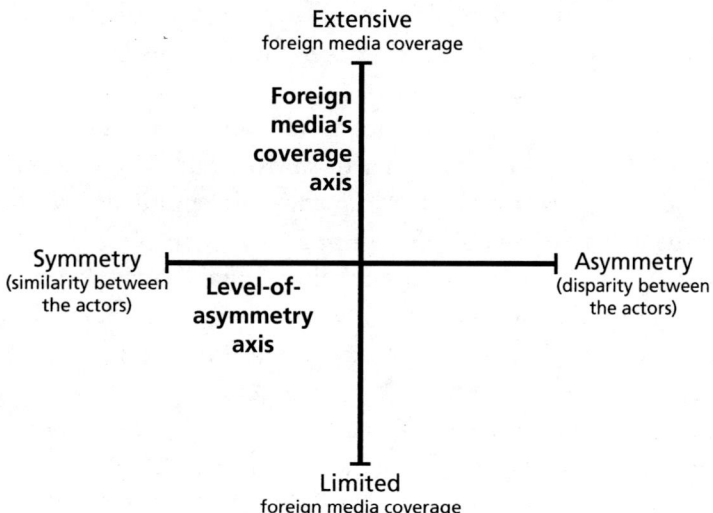

SOURCE: Ayalon, Popovich, and Yarchi, 2014, Figure 1.
RAND *CF334-6.1*

---

[1] This presentation is based on Ayalon, Popovich, and Yarchi, 2014.

These two factors have changed significantly in the past few decades and ultimately influence the outcome of modern conflicts.

The horizontal axis of Figure 6.1 illustrates five asymmetric measures:

- *role in the international system:* measured by the disparity in the international role of the counterparts
- *material symmetry factors:* measured by the disparity of an actor's geography (size of the controlled territory) and demography (size of the controlled population)
- *conflict perception factors:* measured by the disparity in an actor's character-of-conflict perception, the military action logic that needs to be embraced, the chosen policy, the chosen military strategies, the character of the attacked centers of gravity, and other factors
- *hard symmetry factors:* measured by the disparity in, for example, an actor's military capabilities and technology
- *soft symmetry factors:* measured by the disparity in, for example, an actor's economic resources or power, type of regime, the existence of alliances, the consideration given to the international community's opinion or criticism (including international law) on the conflict and its ability to influence the battlefront, the political culture, the dependency on the home front and its ability to influence the battlefront, or an actor's attitudes toward his or her losses (personal and material)

The horizontal axis of Figure 6.1 moves from similarity of power between the actors to a disparity of power between the actors.

The vertical axis of Figure 6.1 measures the amount of foreign media coverage of the conflict on a continuum of limited foreign media coverage to extensive foreign media coverage. Camera and video phones, the Internet, and other technological developments have altered the flow of information around the world and limit states' abilities to control images and messages. Borders are no longer geographic and depend on where people obtain their information.

Moreover, people demand and expect information instantaneously. Ultimately, these factors combine to change how news is processed.

The aim of terror is not the total immediate number of casualties but the exposure to the media (echoing Jenkins' famous quote that "terrorism is theater"). One need only to look at ISIS and its effective use of social media. Filming and uploading videos of beheadings incur minimal cost and provide major psychological impact.

Previously, news dissemination was the domain of established news organizations with set protocols. These organizations acted as "gatekeepers." In the current environment, however, news has a life of its own. ISIS earns status from its followers with its videos—just making threats can gain attention on social media.

Because the media are the principal means through which the majority of individuals receive their information on foreign affairs, media coverage will have a great impact on public opinion, which will subsequently affect policy (similar to the notion of public diplomacy).

Filming and uploading videos of beheadings incur minimal cost and provide major psychological impact.

Applying the asymmetry-of-power and media-coverage axes facilitates explanation and analyses of the differences between the four types of modern conflicts (as shown in Figure 6.2), in which at least one of the adversaries is a state actor.

Figure 6.2
Foreign Media Coverage of Four Types of Modern Conflict

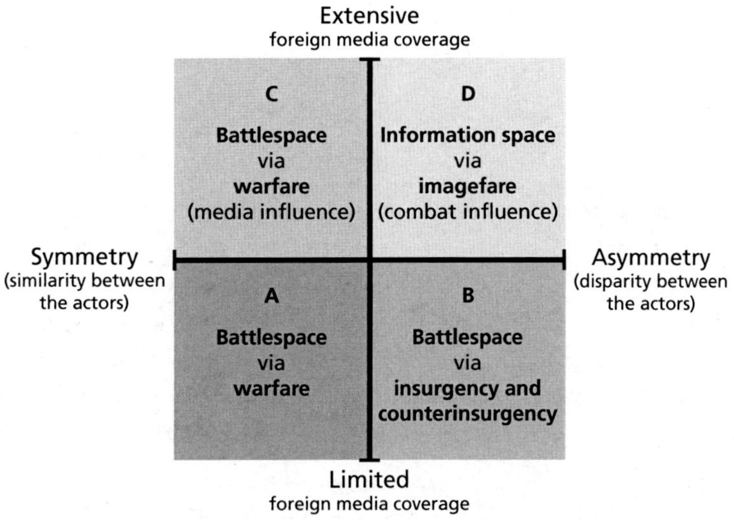

SOURCE: Ayalon, Popovich, and Yarchi, 2014, Figure 2.
RAND *CF334-6.2*

The greater the level of asymmetry between actors (in the factors discussed above) and the more extensive the foreign media coverage of the conflict, the more difficult it is for states to achieve their political goals by using military means alone. As an extreme example, Yarchi suggested the image of a child facing a tank. The media usually will support the underdog and promote the victim.

Figure 6.2 presents four potential quadrants to categorize conflicts in which at least one actor is a nation-state. Quadrant A is characterized by symmetry between the actors and limited media coverage. The traditional approach claims that victory in this kind of conflict will be achieved in the battlespace, and usually through "regular" warfare. When the level of asymmetry increases, however, classical doctrines are no longer valid—even in a conflict between two states. When the asymmetry increases further, particularly when one of the belligerents is a nonstate adversary, the disparity between the actors causes the weaker side to seek to rebalance the conflict through "irregular" warfare.

Quadrant B is characterized by asymmetry between the actors and limited foreign media coverage. Achievements in these types of conflicts occur in the battlespace and are usually through insurgency or counterinsurgency techniques.

Quadrant C is characterized by symmetry between the actors and extensive media coverage. The traditional approach claims that achieving victory in this kind of conflict will occur in the battlespace, and usually through the four fundamental elements of the "Revolution in Military Affairs"—dominant maneuver, precision strike, space power, and information warfare, which also includes managing the media.

Quadrant D is characterized by high asymmetry between the actors (usually, but not limited to, nation-states versus nonstate adversaries) and extensive foreign media coverage. An increasing number of conflicts during the last decade of the 20th century and the beginning of the 21st century can be described as D-type conflicts or events. We argue that quadrant D

Today's conflicts are being fought simultaneously on four different fronts: the military, the diplomatic, the media, and the legal.

conflicts are conflicts occurring in the information space, in which image considerations play a significant role.

Today's conflicts are being fought simultaneously on four different fronts: the military, the diplomatic, the media, and the legal. Perhaps the major change in the modern character of conflict is the influence of media coverage on different political actors' behavior, which affects all other fronts.

Unlike the conflicts with achievements determined in the battlespace in which the main tool for overpowering is warfare, in conflicts with achievement determined in the information space, the actors should first and foremost consider image concerns and use imagefare: *Imagefare is the use, or misuse, of images as a guiding principle or a substitute for traditional military means to achieve political objectives. The actors involved in the conflict attempt to promote their preferred messages through the media in an attempt to gain the public's support and, ultimately, achieve their political goals.*

When actors in a conflict engage in imagefare, compatible with the "theater of terror" but in a much broader sense, they plan their actions according to newsworthiness considerations because the people immediately affected by the event (such as casualties) are not the target—the audience learning about the events through the media is.

With imagefare, each side must show that it is justified in its actions. Yarchi cited the 2010 Turkish Gaza flotilla as an example. Israel stopped six ships from reaching the shores of Gaza. While this was an easy military success, Israel lost in the court of public opinion, suffered a political defeat, and had to endure several international inquiries.

The conclusion is that states recognize the important role that information and images play when they are engaged in a war, but they still fail to use them to their advantage. State actors need to adapt their doctrines once again and realize that modern conflicts have several different (but overlapping) fronts, each with different ground rules and each needing to be won separately. To achieve their political objectives, states involved in conflicts with high levels of asymmetry between actors and extensive foreign media coverage should arguably use imagefare as their primary doctrine because image concerns should lead the decisionmaking process in conflicts in today's information age. In conflicts occurring in the information space, the war is becoming a war of ideas, and victory is achieved in the eyes of spectators worldwide; therefore, states' military (warfare), diplomatic, and legal (lawfare) actions should be guided by imagefare considerations.

Discussion, Led by Todd Helmus

The discussion began with reflections on America's own recent experience. In Iraq, appalling images from Abu Ghraib tarnished America's reputation. The dissemination of these images inspired acts of terrorism and caused foreign fighters to travel to Iraq, making the conflict even bloodier. In Afghanistan, Koran burnings and other inappropriate actions (including some carried out by fanatics in the United States), along with other negative images, undermined America's standing among Muslims and further galvanized its foes. Errant air strikes caused additional damage to America's image.

Unfair or not, in today's world, there is a "tiering" of expectations among combatants. The world holds those it labels terrorists to lower standards of behavior. They may carry out deplorable atrocities and receive only cursory condemnation—after all, they are terrorists, what do you expect? Democracies, on the other hand, are expected to behave as if armed conflict were a genteel debate.

Nobody in the Israeli military cared about images. But the interception of the flotilla was not a naval battle. It was all about images.

It is critical to understand the power of images. American soldiers will soon be videotaped on the battlefield. One terrible image can undermine an entire campaign. When and where incidents occur, governments must be forthcoming. Governments have to learn how to effectively respond to civilian casualties in a transparent manner. Efforts must be made to instill values during military training.

Much discussion revolved around Israel's interdiction of the flotilla sailing from Turkey and Greece in 2010. Ayalon said that understanding the importance of images is critical. He returned to the example of the Gaza flotilla. Nobody in the Israeli military cared about images. But the interception of the flotilla was not a naval battle. It was all about images—a show with spectators, in which Israel was assigned the role of the villain and willingly played the part.

One must see things from the perspective of the audience spectator to assess who is the "good" combatant and who is the "bad" one. In the flotilla incident, the crew and passengers were seen as the underdog—which, in asymmetric warfare, is often the case for nonstate adversaries. This provides an advantage in garnering audience sympathy. The underdog will be perceived as the "good guy"—and the United States and Israel must live with it.

The people on the ship reasoned that, by exposing themselves to danger and death, they would win, and, in the end, they were right. Israel won on the battlefield—the ships did not reach Gaza—but it lost in the political arena and suffered strong international condemnation. *Asymmetry* refers not just to a disparity of weapons but also to a disparity in narratives.

Spirtas pointed out that the United States interdicts ships coming from Mexico all the time. These ships are escorted to San Diego without incident. In the Turkish flotilla, only one of the ships presented a problem. The matter should have been raised above the level of a captain.

Media coverage is now a reality, Yarchi emphasized. States cannot always control the narrative, but they may be able to change it. In the most recent Gaza operation, Israel did its best to prevent civilian deaths but failed to alter the narrative.

Governments must be able to accept responsibility when things go wrong. Diplomats, legal experts, *and public relations advisers* must be involved in the decisionmaking process from the outset. Soldiers must be taught that photos can cause damage.

This prompted Jenkins to remind participants that, while training is important, one has to keep in mind the perspective of a typical young soldier on the battlefield—frightened, full of anger from witnessing the killing or maiming of his comrades, he is not ISIS's knife-wielding thug about to behead a hostage, but he is likely to be concerned more about personal survival than public relations.

Allen Myer agreed that images are powerful and reminded the workshop participants of the iconic 1968 photo of a Vietnamese police chief executing a Vietcong soldier in the middle of a Saigon street. This single photo proved critical in changing America's public perception of the war. *The New York Times* did not fully cover Hamas's tunnels, and this negatively affected people's perceptions of the war because the tunnels became one of the key reasons for the increased Israeli action.

Jenkins said that nonstate adversaries clearly understand the power of images and of the media in general. Lenin claimed that film was the biggest weapon in the Bolshevik arsenal.

Ayalon said that war is all about politics. The law gives a line of defense but cannot overcome the power of images. Israeli artillery fire caused a number of civilian deaths during the recent Gaza operation, which, understandably, caused problems.

Yuval Shany added that it was important not to alienate allies and to maintain international support. He pointed out that, in the example of the flotilla, the Turkish citizens killed were not terrorists. Israel simply did not want the ship to land. Accepting a certain level of risk may be necessary to offset the problems associated with potential negative images.

Another participant noted that the United States strived for precision in its bombing campaigns, but applying military force is not surgery. Using a 500-pound bomb to eliminate a rocket launcher risks collateral damage. Jenkins added that, thus far, civilian casualties resulting from the coalition bombing campaign in Iraq and Syria had not received much international attention—perhaps ISIS's bloody record muffled criticism—but one stray bomb could produce the casualties that would erode resolve to continue the effort.

Tal Mimran said that the Israeli Air Force was proud of "roof knocking," that is, giving occupants of a building suspected of housing enemy arsenals warning that the building is about to be bombed so that they can evacuate. The measure is clearly intended to reduce civilian casualties, but critics abroad still found problems with the practice and condemned Israel for the bombings. In other words, an attempt to reduce civilian casualties did not alter attitudes toward Israel.

Larson argued that it was not simply how a government behaved in battle that counted. While military operations occur from time to time, diplomacy and information warfare are continuous. "Audiences must be developed in between intifadas."

Ayalon reminded participants that Israel was the "good guy" from 1993 to 1999, following the Oslo Accords. It was not simply a matter of image. Israel's policies contributed to its standing. He pointed out that the Palestinians have managed to maintain international sympathy for their cause. Tactical measures to reduce collateral casualties may moderate criticism, but they do not suffice to alter fundamental attitudes. Shany agreed. High-level political decisions set the narrative. "Governments that are seen to behave as oppressive colonial regimes acting against landless people will never be able to successfully tell their side of the story." At the same time, Israeli policies, seen abroad as bellicose, play well with domestic audiences, which ultimately determine who wins elections. Eli Bahar reminded participants that Henry Kissinger observed that Israel has no foreign policy, just domestic politics. It is not easy to satisfy domestic public opinion and world opinion at the same time.

Tactical measures to reduce collateral casualties may moderate criticism, but they do not suffice to alter fundamental attitudes.

Audiences' biases play a role. Despite ISIS's atrocities, which the group itself is determined to advertise, public opinion polls indicate that ISIS continues to attract remarkable levels of support in European countries. Thousands of foreign volunteers have joined ISIS. Knowing what ISIS does—for many, *because* of what it does—thousands more are expected to join. This is an audience apparently beyond the reach of any humanitarian concerns. And how does mediafare deal with audiences, large numbers of whom still believe that the 9/11 attacks were carried out by the Central Intelligence Agency (CIA) or the Mossad?

Taken to its extreme, warfare, to be fair, must involve an equality of suffering. Military commanders and their elected civilian masters would likely disagree.

One Israeli participant commented that reporters themselves are seldom neutral. They have clear biases: some political, others driven by the need for a scoop. Individual decisions are unpredictable. Yarchi related the story of a foreign reporter who was about to cross from Gaza to Israel when she heard about a suicide bombing in Israel. Instead of going to Israel to cover the bombing, she decided to stay in Gaza, anticipating that there would soon be an Israeli reprisal, which she could cover. Her choice between covering a story in which Israelis would be the victims and a story in which Palestinians would be the victims had negative consequences for Israel. The side perceived as the victim has a head start in the mediafare contest.

An undercurrent of complaint flowed beneath the discussion. Many Israelis are convinced that world opinion is inherently biased against them. Some participants attributed this to Israeli policies, and many were openly critical of Israel's tactical mistakes, but they were not entirely convinced that a fundamental change in policy, if that were possible, and a savvier mediafare strategy would alter the deep-rooted hostility that seems to many to prevail. (The same might be said about widespread anti-Americanism.) In addition to all of the other possible explanations for such animosity, legitimate and otherwise, the disadvantage that both nations face is the perception of power, possession of technologically superior arsenals, and prowess on the battlefield, which make any engagement appear to be inherently unfair and attract sympathy for the disadvantaged. Taken to its extreme, warfare, to be fair, must involve an equality of suffering. Military commanders and their elected civilian masters would likely disagree.

CHAPTER SEVEN

Session Five: Lawfare

Opening Remarks by Eli Bahar

The legal dimension of asymmetric warfare is often called "lawfare." This is a relatively new area because the nature of modern warfare has changed dramatically from that of the "classical" wars of the past. In classical warfare, the enemy is visible, and soldiers are easily identifiable by uniform and openly carry weapons. The use of lawfare is part of the larger pursuit of legitimacy.

By contrast, in asymmetric warfare, the enemy is usually invisible, hiding among the civilian population, often in densely populated areas. Lethal attacks are often launched from civilian facilities. There may be no means to distinguish combatants from the civilian population.

In classical warfare, a democratic nation's obligation and responsibility is to conduct the war according to the rules of war—especially the principle of distinction between combatants and civilians.

When it comes to asymmetric warfare, the opponent often targets civilians, not only ignoring the rules of war but deliberately doing so as part of an overall strategy against the democratic state.

In classical warfare, the ultimate goal of both sides is to defeat the enemy with respect to its capabilities to use military power. In asymmetric warfare, the opponent's goal is not to defeat the state's armed forces but rather to make the civil society so terrified and concerned that it will pressure politicians to withdraw from the state's positions or abandon its policy aims, thereby losing the war not through the battlefield but through determination of the democracy not to continue fighting.

In classical warfare, the territorial and temporal limits of the conflict are relatively defined. The nature of asymmetric warfare is much more amorphous. It is not limited to a certain territory or distinct timeline.

In classical warfare, the enemy's fighters are essentially anonymous: It is not important to know the name of the enemy's soldier or commander before attacking him.

When it comes to asymmetric warfare, in many cases, it is crucial to know the opponent and to have very personal and detailed information as a precondition for determining the legitimacy of a strike.

These new features are different from the set of assumptions that were the

> *When it comes to asymmetric warfare, the opponent often targets civilians, not only ignoring the rules of war but deliberately doing so as part of an overall strategy.*

basis for the laws of war, especially international humanitarian law. These laws are the basis of the legal norms, binding democratic nations to conduct their military power accordingly.

This differentiation, combined with the emerging power and influence of international tribunals, is known as lawfare.

Lawfare is often used as a negative term, suggesting manipulation, although it is not limited to that. Ironically, it is an area in which the democratic state and its officials feel vulnerable. In contrast, the opponent often deliberately violates the norms while simultaneously using them to weaken democratic nations. Lawfare is used to counter the weapons of the democratic state by exploiting its own laws and judicial systems. It focuses on government and personal liability.

Because asymmetrical warfare takes place in densely populated areas, it inevitably generates more grounds for legal action. While nonstate adversaries typically do not comply with international humanitarian law, as already noted, they will simultaneously use that law to undermine the motivation and legitimacy of their democratic state foes.

A number of Israeli leaders are subject to universal jurisdiction. For example, Ayalon cannot travel to the Netherlands or Spain without being subjected to arrest. Other political and security leaders, including former Shin Bet director Avi Dichter, former foreign minister Tzipi Livni, and former Israel Defense Forces (IDF) commander Doron Almog, are currently subject to universal jurisdiction claims. International tribunals, such as the International Criminal Court, are used against Israelis. An attempt was made to bring David Benjamin, a reserve IDF officer and South African native, to trial based on his legal advice prior and during the 2008 "Cast Lead" operation. Domestic courts could also be used as part of lawfare.

There are inherent difficulties in applying the norms of international humanitarian law to asymmetric warfare. Applying the concept of *proportionality* is next to impossible and provides no guidance to the commanders on the ground, since it comes without clear guidelines. It is also difficult to apply the fundamental principle of *distinction* in a civilian environment, since the entire battlefield is often a civilian area, making it nearly impossible to distinguish between combatants and civilians. Moreover, the application of the principle of *military necessity* is problematic when it relies heavily on intelligence and other secret evidence. Finally, asymmetric warfare presents challenges to efficiently striking the opponent without violating the principle of *perfidy*. Effective warning—for example, roof knocking—weakens the chances of a successful military mission and places soldiers at additional risk. Other challenges include striking political or religious targets, regardless of whether they are also being used to support active military operations. What about individuals like Hamas leader Ismail Haniye? Or how should armies treat bridges or electricity? In World War II, such targets were bombed for obvious military reasons. In asymmetric warfare, they may be considered civilian facilities.

Consequently, the challenge is how to adjust international humanitarian law to apply to modern asymmetric wars. There needs to be legal recognition of the constant state and timeless nature of armed conflict against nonstate adversaries. There needs to be legal application of the principle of self-defense against nonstate adversaries when there is no other alternative (for example, in failed states). Given that liberal democracies are often on the front

> *Lawfare is used to counter the weapons of the democratic state by exploiting its own laws and judicial systems. It focuses on government and personal liability.*

[T]he law should be the same for all belligerents in all conflicts, but, in reality, it is applied unequally to state and nonstate adversaries.

line in the fight against nonstate adversaries, there needs to be a flow of information among them as well.

Furthermore, it is necessary to acknowledge the role that intelligence plays in winning the war and to determine the "military necessity" and the key element of intelligence for the principle of *distinction.* Also, wider legal tools and wider public control over intelligence agencies should be considered.

It is important to develop a publicly available code of conduct for certain military actions, such as targeted killings. It is important to know who is making the decisions and under what guidelines and circumstances targeted killings are allowed. Sharing more intelligence publicly assists in efforts to win the imagefare battle. The final point Bahar made was on *proportionality*, and he wondered whether a different application of proportionality was possible between "limited" nonstate adversaries and "total" state adversaries.

Discussion, Led by George Jameson

Lawfare can be used both positively and negatively. The United States prides itself on adherence to a system of laws, rather than morality or culture, to justify its actions. But what do nonstate adversaries think? Lawyers have many domestic and international laws to consider, including the law of armed conflict and human rights law. All of these laws form a basis for a rational discussion.

Certain activists like to argue for the application of human rights law, in which the use of force is a last resort. In contrast, under the law of armed conflict or humanitarian law, the use of force can be a first resort, but it must be proportional. Much in the debates over the application of law depends on the type of lawyers involved. For example, work on the Yugoslav war crime trials was carried out by human rights lawyers, not law-of-war lawyers.

Another problem dates back to the Treaty of Westphalia in 1648 and deals with what happens among states rather than within a nation-state. Some countries, such as Syria, China, and Russia, argue that what happens within their borders is an internal matter and therefore not subject to international external interference. This situation arises not only during the conduct of war. Around the world, cultural views of the laws and their limits differ. In the United States, the thinking and legal approach on some issues goes back to the 1800s (requiring a "shocks the conscience" standard in apprehending felons, for example), while Europeans have their own notions of privacy, warrants, and other issues. In the United States, the government is more restrained in some areas, and private entities, such as Google, collect a lot more information than the government does. The approach in Europe is very different, enabling greater government leeway and more constraints on private corporate conduct. Law is mobilized to bolster these different national and cultural ideas.

Part of the discussion focused on the fact that the law should be the same for all belligerents in all conflicts, but, in reality, it is applied unequally to state and nonstate adversaries. Shany pointed out that this is a battle that the state will always lose. Nonstate actors operate with immunity, since they often have nothing to lose. Johnson also questioned whether it is law when one side follows it and the other side does not. Law legitimizes actions, but *legitimacy* is a slippery term. Does it mean anything goes if the cause is just? Who defines what is just?

Jenkins pointed out that this was the original problem in defining terrorism in the 1970s. Some argued that those fighting against colonial and imperialist governments in wars of national liberation could not be called terrorists no matter what they did, because their cause was just. The term "war of national liberation" is no longer used, but the concept has survived.

Brennan noted that ISIS is a quasi-state, and the international system is based on state actors. (Participants returned to this topic in the following discussion. Some expressed the view that the so-called Islamic State is a pretension. True, it controls some territory and population, but so do many insurgencies. ISIS has no identifiable government other than a man who asserts that he is the caliph and therefore the ruler of all Islam. The organization has no real functioning economy. Why should it be treated as a state? And what difference would it make?) Some participants thought it was time to think about changing the international legal system, but Mimran observed that changing that system would be difficult and might ultimately require another world war.

There was also the general view that constraints are increasing. The Geneva Conventions are supposed to govern warfare, but asymmetric warfare remains largely unregulated. In asymmetric war, human rights law is more developed. It errs on the side of caution, with more rigid rules of proportionality. Obama has set an even higher standard to govern the use of drones, but it still does not satisfy those who claim that such attacks violate international law. The Israel Supreme Court has said that roof knocking is legitimate. Others say it violates the rules. Some urge erring on the side of caution.

The discussion of giving specific warning before air attacks brought up the issue of human shields. In Serbia, civilians gathered on bridges in Belgrade to prevent them from being bombed. In the Gaza operations, roof knocking led to evacuations, but, in some cases, it also persuaded occupants to gather on the roofs to prevent bombing. (Israeli authorities claimed that Hamas instructed occupants to go to the roof.) Law prohibits the deliberate use of human shields, but locating weapons inside of or adjacent to civilian facilities and exploiting human shields have become part of nonstate adversary tactics.

The rules of war were developed after World War II. One participant asserted that, under current international law, President Harry Truman would be treated as a war criminal. Dropping the atomic bomb on Japan could not be shown to have resulted from military necessity. (Ordinarily, this would be hotly debated, but it was allowed to pass unremarked.) Another observed that the United States did not give specific advance warnings before striking targets during its bombing campaign, but the Israelis do. Another pointed out that U.S. military actions during the Kosovo War were determined by the military commanders—they were the ones who defined what was a military necessity, not lawyers reviewing the case after the fact.

> The Geneva Conventions are supposed to govern warfare, but asymmetric warfare remains largely unregulated. In asymmetric war, human rights law is more developed.

Jameson added the comment that one of the factors driving change is the changing definition of combatants. In some cases—for example, special operations to capture terrorist leaders or targeted killings—it is not clear if the combatants are engaged in a war or in law enforcement. The United States has gone back and forth between the two, but it is generally accepted that both can apply.

Jenkins agreed that counterterrorism was blurring war and law enforcement but argued that it was creating new problems. Some participants asserted that, although not yet codified as law, certain military actions must have judicial authorization. The battlefield is becoming a "crime scene." The United States had partially accepted this when it established a rough equivalent of due process for targeted killings, claiming that, during war, it is legitimate to target enemy commanders. And if action on the battlefield is a component of law enforcement, does that change thinking about civilian casualties?

The problems go both ways, Jenkins pointed out. The U.S. Senate passed legislation authorizing the president to arrest and indefinitely detain terrorist *suspects* without habeus corpus, arguing that there was no difference between an enemy combatant captured on the battlefield and a terrorist operative captured in the United States. The legislation was subsequently amended, but the idea of treating all terrorists, including suspects, as enemy combatants continues to have powerful proponents. This, in Jenkins' view, is a direct threat to democracy.

It was also noted that some legal scholars are trying to introduce a "fairness doctrine" into warfare, arguing that certain weapons and tactics should be banned on the grounds that they do not expose the operators to the same risks as the target and therefore are inherently unfair. This could apply to the use of airpower against adversaries who lack aircraft, but it could also be extended to any other heavy or remotely operated weapons. Lawfare is what you use when you cannot compete militarily, noted Johnson.

One person expressed the view that lawyers get in the way, and the participants discussed the role of the lawyer in light of the power of the commanders in the field and the role of policymakers in making the laws that must be observed.

Jameson warned against the legal tail wagging the dog. Laws are not designed simply because they are what lawyers want. That said, the values and principles expressed in law are important. They do not win or lose battles, but they are part of winning wars—they are one facet of the conflict. Ayalon noted that lawfare provides a basis on which to avoid losing identity as a democracy. There are human and moral principles to consider, as well as principles of economy. Moreover, government and military commanders act in a certain manner because upholding these values is in their own interest. Beer added that most laws help soldiers, and there are good reasons for these rules. As an example, the law of *military necessity* coincides with the principle of *economy of force*.

Who should investigate violations of the rules: military courts? civilian judges? international tribunals? Shany challenged the quality of the Israeli investigations into crimes allegedly committed by Israeli soldiers. There is no transparency and no independence in these investigations. The military decides the outcome of the cases. In one of the few cases prosecuted, an Israeli soldier was convicted of stealing a credit card in the "Cast Lead" operation. Shany added that the United States was no better when it comes to individual accountability. Another participant noted that the United States had not signed on to the International Criminal Court.

Ayalon took a different view. He noted that, in his 33 years in the military and the Shin Bet, he did not always

> [G]overnment and military commanders act in a certain manner because upholding these values is in their own interest.

The battlefield is not the main theater of the war against terror.

admire the Israeli legal system, but he has since learned to admire the Israeli Supreme Court. The actions of the Israeli Supreme Court are unique in the world. Like no other court, it has asserted its primacy in deciding and framing the boundaries of acceptable tactics. It has ruled on human shields and destroying houses. In December 1999, the court said that Israeli authorities could not continue interrogating prisoners in the same manner. Initially, Ayalon felt that this decision forced him to operate with one hand tied behind his back, but he has learned to admire it. He is now convinced that the way Israel was acting posed a threat to its own democracy. No other supreme court makes the types of decisions that are reviewed by the Israeli Supreme Court.

Ayalon went on to argue that it is the absence of strategy, not the application of law, that is the problem. The battlefield is not the main theater of the war against terror. Israel often acts out of military necessity, but too often without strategy. Recalling his experience as the director of Shin Bet during the second intifada, Ayalon said that, in his view, the problem was not one of law; the problem was Israel's strategy, which only created more enemies.

The subject of negotiations arose briefly during this session. An American participant asked whether there are any legal limits on negotiating with terrorists. Jenkins said, No. Public statements declare that the U.S. government will not make concessions to terrorists holding hostages, but this is policy—a guideline to action—not statute. When policy proves counterproductive, it will be set aside. Nothing in the policy precludes communications, even with terrorists. The United States has not interfered with ransom payments by private parties. Policy does not prevent negotiations to end conflicts or exchange prisoners.

CHAPTER EIGHT

Session Six: Political and Diplomatic Dimensions of Conflict— Lessons from Around the Globe

Opening Remarks by James Dobbins and Rick Brennan

Dobbins began his discussion by contrasting recent conflicts fought by Israel with those fought by the United States. The United States almost always fights in international coalitions, while Israel fights alone. As an example of coalition fighting, Dobbins cited the coalition of countries involved in the air campaign against ISIS. In addition, the United States usually has the benefit of fighting with a United Nations resolution and/or a congressional mandate. In the Bosnia and Kosovo ground wars and in Afghanistan (but not in Iraq), there were United Nations resolutions approving the use of military force. These provide some degree of political coverage for military action. Israel's status in international arenas has often been much more tenuous, further constraining its already limited range of actions.

In the case of Israel, the domestic population is more exposed to harm and clearly at greater risk than that of the United States. Not surprisingly, domestic pressure for taking whatever military action is deemed necessary to reduce the threat is greater in Israel than in the United States. Normally, the American public is more insulated from attack, and that offers some degree of insulation from domestic political pressure. The 9/11 attacks were an exception and provoked a prompt military response.

Given that the United States usually operates in coalitions, American diplomats play an important role in setting the parameters for military action. This is not the case in Israel.

Further, most of the asymmetric conflicts in which the United States has engaged have been at the invitation of and in partnership with host governments, offering a very different range of challenges and pitfalls from those faced by Israel in its unique situation. This is true of the U.S. hoped-for end states as well. U.S. goals, however vague and shifting, often involve increasing the capacity and stability of host-nation governments. In contrast, Israel has, arguably, been thus far unable to clearly and coherently define a desirable end state and work toward it. For many reasons, this is more difficult in the case of Israel's troubled relations with the Palestinians, but it also reflects domestic Israeli attitudes and politics. The contrast is evident in the terms used. The United States engages in nation-building—a concrete, if difficult, goal, while the Israelis engage in what has been described as "mowing the lawn," implying that it is an endless task.

Terminating conflicts in the Middle East with a negotiated settlement is never easy. Given U.S. reluctance to negotiate directly with terrorists, negotiations are

> *Israel has, arguably, been thus far unable to clearly and coherently define a desirable end state and work toward it.*

often conducted through proxies or patrons. In the case of Israel, negotiations have been conducted with Hamas and Hezbollah to end hostilities, as well as to return hostages and bodies, but without conclusive peace negotiations.

The United States has also had some success in creating alliances with local nonstate actors. American support for more-moderate insurgents has limited the role of extremists in several situations. The conflict in Kosovo and the Kosovo Liberation Army offer an example. Once the United States and the North Atlantic Treaty Organization (NATO) provided support to the Muslims fighting in Kosovo against the Yugoslav security forces and Serb paramilitaries, the Kosovar extremists were marginalized. The removal of the Taliban government in Afghanistan was facilitated by American cooperation with the Northern Alliance in 2001. The support offered to and assistance provided by the Sunni tribes in Iraq in 2006 was critical to the successful campaign waged against the insurgents. In post-Qaddafi Libya, the United States had the opportunity to support the moderate opposition at the expense of extremists but failed to do so.

America's experience offers several lessons:

- Supporting insurgents—for example, in Syria—may, in some cases, be desirable. This is not always possible. Sometimes insurgencies work against U.S. interests.
- Governments must identify and pursue an end state. It is critical when developing strategy, even though end states are not always achieved. The United States did not achieve an end state in Iraq by the time it withdrew its forces in 2011. Afghanistan may present the same problem. For Israel, since it has not set an end state, it is impossible to reach this goal.
- Governments should seek opportunities for reconciliation as is currently being done in Afghanistan. Since 2011, the United States has been seeking to work with the Taliban—both directly and indirectly—without preconditions. The United States even exchanged prisoners with the Taliban, although this caused considerable controversy. The United States approached negotiations with Iran on a similar no-preconditions basis. Once the nuclear issues with Iran are tackled, other agenda items, such as resolving the conflicts in Afghanistan (discussed in the past) and Iraq, can be addressed. Again, this is not without controversy. However, the United States has parallel interests with Iran in Iraq. Presently, nobody in Israel or the United States (largely in deference to Israel) will officially talk to Hamas. Hamas is an insurgency and not just a terrorist group, so there is little hope of defeating it or ending the conflict without negotiations.

Dobbins' final observations raised the question, How does a nation get out of a war? Brennan focused his remarks on RAND's research regarding the withdrawal of U.S. troops and the end of U.S. military involvement in Iraq and the lessons learned for ending future wars (Brennan et al., 2013). When the U.S. military formally ended armed operations in Iraq on December 18, 2011, it was the toughest departure in U.S. history. In addition to the 100,000 military and civilian personnel that would require redeployment, the United States had during the Iraq War deployed millions of pieces of equipment, while the U.S. military had managed hundreds of bases and other facilities. An already complex situation was made more difficult by the fact that no one was certain whether there would be a residual U.S. military presence after 2011 or, if there were such a presence, its extent. The American departure from Iraq also illustrated some broader issues about the relationship between military action and political achievement.

In planning a military campaign, the initial focus is on achieving the desired military objectives. Minimal consideration is given to how the military action will achieve the stated political goals. As the costs of the war (in terms of lives lost, financial costs, and debilitating distractions from other matters) inevitably increase, these political goals tend to expand. Wars follow their own logic, which is not predictable at the onset.

When policymakers decide to go to war, they must understand that they lose control the minute they enter the conflict.

When policymakers decide to go to war, they must understand that they lose control the minute they enter the conflict. What began as an interstate conflict when the United States invaded Iraq quickly evolved into a counterinsurgency campaign, for which the United States was ill-prepared.

Mounting costs in the absence of visible progress erode domestic political support and dissolve the initial political bipartisan support for the continuing effort. Political leaders confront the hard choice of backing down while still claiming some measure of success or intensifying the war effort to justify the investment with something that can be portrayed as victory.

The United States faced such a calculus in 2005 and 2006 as the situation in Iraq deteriorated. In that case, the president supported a surge in U.S. forces. The surge blunted the immediate threat but made the United States more anxious to withdraw. The Bush administration wanted no part of a civil war. The succeeding Obama administration was anxious to end American involvement in a war started by the previous administration and was committed to getting the military out of Iraq as soon as possible. In hindsight, assuring reconciliation should have been a higher priority, but the underlying current was that the United States is not an imperial power and wanted to go home.

The take-away from the Iraq analysis is that "war termination planning must start before the war begins."

To underscore the point, Brennan offered a quote from Winston Churchill's book, published in 1930, *My Early Life: A Roving Commission*:

> Let us learn our lessons. Never, never, never believe any war will be smooth and easy, or that anyone who embarks on that strange voyage can measure the tides and hurricanes he will encounter. The Statesman who yields to war fever must realize that once the signal is given, he is no longer the master of policy but the slave of unforeseeable and uncontrollable events. Antiquated War Offices, weak, incompetent or arrogant Commanders, untrustworthy allies, hostile neutrals, malignant Fortune, ugly surprises, awful miscalculations—all take their seats at the Council Board on the morrow of a declaration of war. Always remember, however sure you are that you can easily win, that there would not be a war if the other man did not think he also had a chance. (Churchill, 1930)

The shift in thinking was reflected in the 2009 Joint Campaign Plan. Its stated objective was to produce "a long-term and enduring strategic partnership between the United States and a sovereign, stable and self-reliant Iraq that contributes to the peace and security of the region." This required the development of legitimate and participatory governance (a political goal),

The Iraqi government did not help resettle returning refugees, thereby creating a potential breeding ground for a resurgent Sunni insurgency.

an army and police that would be able to ensure internal and external defense (a security goal), a diverse economy (an economic goal), and a sovereign, self-reliant country that contributes to peace and stability in the region and beyond (a diplomatic goal).

In fact, about 20 percent of the campaign goals were military, and the remaining 80 percent were focused on building a viable and effective government. The goals were correct, but there was little progress. Government capacity was improved, but this did not lead to an inclusive government.

It became increasingly clear that a residual force would be needed to maintain order and protect the Sunnis from repression. Nonetheless, the American departure continued, and political goals remained unmet, in large part because of the existing ethno-sectarian divisions and hostile regional interests at work in Iraq.

Negotiators made limited progress in easing Arab-Kurdish tensions, especially over sharing oil revenue, but sectarian tensions between Shias and Sunnis remained. Fearing a Ba'athist resurgence, the ruling Shia majority continued to carry out discriminatory policies, and while "the Sons of Iraq" (a coalition of former Iraqi military officers and Sunni tribal sheikhs) had been theoretically integrated, the government of Iraq remained evasive about any kind of genuine reconciliation. The Iraqi government did not help resettle returning refugees, thereby creating a potential breeding ground for a resurgent Sunni insurgency. Meanwhile, Shia groups were increasingly operating under the control of Iran. Reconciliation came within the Shia community—not between the Shias and the Sunnis.

Similarly, the political landscape in the region and internationally was hardly benign. Russia and China worked against the United States in Iraq. Regionally, neighboring countries conspired to keep the Iraqi government weak. For example, the Saudis continue to have poor relations with Baghdad due to perceived Iranian influence and the broader Sunni/Shia schism. Syria still harbors Iraqi Ba'athists and facilitates the flow of foreign fighters heading to the Islamic State.

The U.S. military's Judge Advocate General and the U.S. Justice Department worked to build an Iraqi justice system, both civil and criminal, but again, sectarian loyalties undermined the effort. Political influence also continues to affect the legal system and its institutions, and de-Ba'athification continues to be a tool used by Shia leaders to target political opponents or seek revenge.

Corruption remains rampant. Iraq was ranked 176th out of 180 countries in Transparency International's 2009 Corruption Perceptions Index—only Sudan, Myanmar, and Afghanistan ranked worse (Transparency International, 2009). This endemic corruption undermines the legitimacy of the government and stifles economic development and the necessary private investment. The collapse of the Iraqi army in the face of the ISIS assault in 2014 could be attributed to cronyism, which put incompetent commanders in charge, but it was also a result of the high-level corruption represented by salaries being paid to nonexistent soldiers.

Discussion, Led by Amichay Ayalon

The importance of international coalitions was a particularly resonant point from the Israeli perspective, as Israel has almost always fought its battles alone.

Ayalon agreed with the conclusion that the United States and Israel are fighting a different type of war, which must be waged on multiple fronts, and that building broad support among the community of nations is a critical component of today's conflicts. That means complying with international legal norms and effective diplomacy.

Diplomacy and military operations are no longer well-coordinated in Israel, according to one Israeli participant who criticized the gaps in Israel's current civilian leadership. He recalled that, in 2006 and 2007, there had been extensive collaboration between Israel's diplomats and military commanders, leading to a joint campaign plan, which ambassadors and military would implement.

The importance of international coalitions was a particularly resonant point from the Israeli perspective, as Israel has almost always fought its battles alone. Indeed, there is not one case of a coalition, going back to ancient history, although one workshop participant suggested that Israel's invasion of Egypt in 1956 was an example of Israeli involvement in a coalition with France and the United Kingdom. The Israeli operation was successful, but the combined effort fell apart in the face of U.S. opposition. The view that Israel nearly always goes it alone was challenged by other participants, who noted that the United States has always provided Israel with strong political and material support.

The political challenges of coalition-building and the nation-specific rules of engagement can be complicated in other ways as well. Some participants noted that creation of a coalition comes with limitations in the form of varying rules of engagement, formal or informal. However, others pointed out that these factors need not necessarily result in the dilution of the effort down to the capability of the most constrained partner, but rather the parceling of missions such that individual partner assignments are compatible with their specific constraints. Practically, this often means that the United States contributes nearly all of the brute military capacity, while the coalition partners may provide political support. Much will depend on which issues are most critical to which partners. For example, in the cases of Bosnia and Libya, European concerns were more directly affected, and therefore the United States played a more supporting role. Varying levels and different types of capacity and concern can thus be complementary rather than just constraining.

The anti-ISIS coalition is an example of the United States building a coalition of partners. The campaign is supported by Jordan and Saudi Arabia.[1] Some European nations joined the bombing effort, while others were willing to provide training. Australia, Canada, and the United Kingdom have limited capacity compared with the United States, but they are well integrated into the air campaign. By allowing participation according to willingness to participate and capability, the anti-ISIS coalition was able to grow.

---

[1] Following the brutal burning to death of a Jordanian pilot, Jordan stepped up its participation in the bombing. Since the workshop, Egypt also bombed ISIS allies in Libya following terrorist attacks on Coptic Christians in Egypt.

[T]he United States and Israel do not fully understand all the ramifications of the Arab Spring, but it is clear that all governments are becoming weaker, and the "street"—the reflection of popular views—is growing stronger.

All participants easily agreed that the Middle East is a messy place right now. The regional landscape and geopolitical alignments are rapidly changing, making it especially difficult to sort out who your allies are or should be in the Middle East, given the various crosscurrents: authoritarian regimes versus more democratic regimes, Sunni versus Shia, ISIS versus everybody. It is like a six-dimensional chess match. Moreover, the sides change all the time. The United States and Israel work alongside last year's enemies against last year's allies. Both countries make judgments as to which foe poses the greater threat.

Brennan observed that Iran was responsible for approximately half the American casualities in Iraq. The Iranians have expanded the conflict in Syria with their own forces and by using Hezbollah. Today, American aircraft support Shia militias that only two years ago were killing Americans. The largest American base in the Middle East is in Qatar, which, the Israeli participants pointed out, supports Sunni extremist groups.

This discussion brought Dobbins back to the issue of a U.S.-Iran rapprochement. The American view is that there are areas of overlapping interests, as well as differing interests, between the two nations. Iran is neither open nor truly democratic, but it is a cosmopolitan country with a well-educated population, and it is more open than most other nations in the region. There is hope that it will move in a moderate direction.

The situation in Afghanistan is probably the greatest common interest of Iran and the United States, and policy differences are minimal. After 2001, Iran supported the coalition to overcome the electoral impasse in Afghanistan. Moreover, the Shia population in Afghanistan is minimal. In contrast, the Israeli-Palestinian conflict remains an area of differing interests. If there is a breakthrough on the nuclear negotiations with Iran, more discussion on other matters will be possible.

Ayalon said that Israel must prioritize threats. The Shia/Sunni division is the dominant conflict in the region at the moment, but it is complicated by regional contests for power. Three countries are vying for dominance in the Middle East: Turkey, Iran, and Egypt. Saudi Arabia has never tried to gain dominance in the region, and Ayalon said he did not see Iran as a critical threat to Israel, although many Israelis would disagree. And while Qatar is not Israel's greatest ally, it is not Israel's worst enemy.

Ayalon went on to observe that the United States and Israel do not fully understand all the ramifications of the Arab Spring, but it is clear that all governments are becoming weaker, and the "street"—the reflection of popular views—is growing stronger. Emerging communication technologies fuel this trend, but so does American policy. "America believes democracy cures everything," observed Ayalon, "so the Palestinians have elections, but then the Americans are surprised when Hamas wins."

But what do these changes mean for diplomacy? In the past, there was an underlying assumption that governments have the authority to make treaties. But now—to an ever-increasing degree—people matter, but public opinion is volatile and can prevent governments from making strategically sensible decisions. Under these new circumstances, would treaties with

Egypt or Jordan still be possible? This development also applies to Israel. In the past, Prime Ministers Yitzhak Rabin or Ariel Sharon could dictate policy, but that is no longer the case.

Although it was not a specific topic of any of the sessions, criticism of the American invasion of Iraq proved to be a recurring theme at the workshop. The criticisms were not confined to the Israeli participants. Dobbins stated that the decision to go to war in Iraq was taken against the advice of many allies and all of Iraq's neighbors, with the exception of Kuwait. Moreover, it was poorly planned and underresourced. In recent years, the United States has intervened in six countries. In each case, indigenous institutions for public safety had been either destroyed or thoroughly discredited by reason of the insurgents' behavior. Security would therefore need to be provided by an intervening power. By the latter years of the Clinton administration, these lessons had been learned—60,000 troops were deployed in Bosnia and 50,000 in Kosovo. It was wishful thinking that Iraq, a nation of 25 million people, could have been effectively policed by 30,000 troops at the end of the year, which is what the war plan anticipated. The Army Chief of Staff, General Eric Shinseki, warned that hundreds of thousands of troops would be required to occupy Iraq but was ignored.

CHAPTER NINE

Session Seven: A Review of the Initial Findings with Invited Guests

Discussion, Led by Amichay Ayalon and Brian Michael Jenkins

Ayalon and Jenkins both wanted an opportunity to engage government officials and solicit their feedback. A number of American government officials were invited to attend the second day's sessions as observers and commentators. In order to allow free discussion, they were promised a strict application of Chatham House rules—no one would be quoted. Session Seven was created to provide an opportunity to recap the discussions before transitioning to the final session on future work.

After welcoming the guests, Jenkins led off with a brief review. It was the aim of this workshop to begin to explore the unique challenges of asymmetric warfare. Apart from its two engagements in the Balkans during the 1990s, American engagements in asymmetric warfare had become mostly a matter of counterterrorism and consequent unplanned counterinsurgency missions in Afghanistan and Iraq. These efforts have become increasingly preemptive and preventive. Jenkins spoke of the persistence of the terrorist threat and the long duration of these wars, which increased the burden of public expenditures and stretched public patience. He cautioned that America's military superiority has not guaranteed success, although the absence of any major terrorist attacks on American soil since 9/11 could be interpreted as a positive result.

Jenkins then summarized the preceding discussion on strategy, the importance of discriminating between different nonstate adversaries, the need for any effort to be seen as legitimate, the growing importance of perceptions, and the social implications arising from the sense of perpetual threat and its potential to erode basic liberties.

He ended with some of the questions faced by American leaders today:

- What role does the United States want and think it should play in the world?
- Is America willing to behave as an imperial power—to fight every potential aggressor on every distant frontier, theoretically in order to defend the American homeland? Or can the American people be persuaded to accept some quantum of risk at home to avoid continuous engagement in conflicts abroad?
- Asymmetric conflicts have been the norm for 25 years. Will they continue to dominate the military challenges faced

Is America willing to behave as an imperial power—to fight every potential aggressor on every distant frontier, theoretically in order to defend the American homeland?

by the United States? Does it make sense to formulate a grand strategy for dealing with the diverse adversaries and circumstances lumped under the label of terrorism or asymmetric warfare?

The United States is still being jerked around by 9/11, the Arab Spring, the energy revolution, events in Ukraine, China's territorial claims, and the Islamic State.

- In many cases, the world expects the United States to act because it can. How can the United States avoid sending American troops into battle while not undermining allies' confidence in American assurances of assistance if they are threatened?
- Can a liberal democracy deal with a perpetual threat, or at least the perception of a perpetual threat, without gradually eroding its own commitment to democracy as extraordinary measures taken to defend the country during times of national emergency become permanent features of the political landscape?

Ayalon added his own observations. Citing Clausewitz, Ayalon said that the nature of war does not change, although its character changes. The past 40 years have seen changes in weapon technology, but even more significant changes have occurred in the field of communications. The Internet and mobile phones have altered the terrain of war. The United States and Israel are entering a new era with the evolution of media and the cultural aspects of globalization. Will it take decades to fully understand the influence these changes have on diplomacy and war?

Today's battlefield encompasses multiple fronts, which must be simultaneously addressed: military, media, diplomatic, and legal. The enemy understands this terrain better than government does. The concurrent technological and cultural revolutions have changed the terrain on all four fronts. For example, the introduction of the miniature camera has allowed disputes at checkpoints to escalate into major international incidents. In Ayalon's view, Israel has lost control of the narrative and its accompanying diplomacy.

Much of the subsequent discussion focused on the issue of America's strategy. Richard Solomon asked, What is America's role in the world? Is there a grand strategy? What do the United States and Israel see today? The Middle East remains in chaos. American efforts in Iraq and Syria remain fraught with uncertainty. The United States is still being jerked around[1] by 9/11, the Arab Spring, the energy revolution, events in Ukraine, China's territorial claims, and the Islamic State. "Jerked around" seems to be the operative phrase. The perception of threats and the national agenda keep changing.

There is equal uncertainty at home. Americans continue to wobble between engagement and isolationism. Senior leaders are expected to articulate and uphold American values while simultaneously avoiding conflicts. But how can America protect Syrian protesters against brutal repression, prevent the massacre of Yazidis, contain ISIS fanatics, and rescue American citizens or respond to their beheadings without military engagement? Still, people unrealistically expect military operations without casualties. Many have become disenchanted with efforts

---

[1]   In this instance, "jerked around" means being confronted with the emergence of multiple security challenges on diverse fronts, making it hard to focus, prioritize, or develop an integrated grand strategy.

at nation-building, arguing that rebuilding America itself must take precedence. A quarter-century after the fall of the Soviet Union, America's national-security policymaking institutions remain stuck in the Cold War.

Ten months before Pearl Harbor, President Franklin Roosevelt outlined his iconic four freedoms that would guide and justify U.S. actions in the international arena. What is their equivalent today?

Perhaps Americans will be the stewards of a Ken Waltz world in which anarchy is the organizing principle.

One of the government participants questioned whether the United States does not, in fact, have a grand strategy. Yes, America often acts too slowly. Every event in the world may not demand an American response. America's strategy for dealing with ISIS has taken time to evolve, but the United States is now leading an international coalition that is reducing ISIS's capabilities and is working with Turkey on refugee issues. These achievements required quid pro quos. Is there a general theoretical framework? There is an opportunity for U.S. leadership. The current administration believes in international institutions, and many of these institutions were created in the image of the United States. Perhaps Americans will be the stewards of a Ken Waltz world in which anarchy is the organizing principle (Waltz, 1959, 1979).

One government guest suggested that the United States and Israel had seen three critical inflection points in recent history, which he identified as the end of the Cold War, the 9/11 attacks, and the present situation. He remarked that it seemed as if the entire framework for apprehending international politics had gone away and had not yet been replaced. He added that perhaps the United States and Israel overlearned the lessons of 9/11 and were still trying to cram something confusingly similar yet distinct into an al Qaeda template. Other guests from government also concurred on the importance of disaggregating the different challenges, agreeing that ISIS's strategies and its effects were very different from those of al Qaeda, although the United States seemed to be stuck in an al Qaeda paradigm.

Guests also noted the role of visionary leadership in these conflicts—leaders who were able to articulate not just threats but also the need for sacrifices and, most of all, who were able to frame public expectations and communicate which entanglements were better avoided and why.

When is military intervention warranted? Given the distance and removal of some conflicts from the United States, some wondered why Americans are always expected to take the lead in issues that do not directly affect their own security. For some, this is a matter of force projection and power vacuums. For others, it involves the protection and assurance of allies. In cases such as the 9/11 attacks or even the beheadings of hostages by terrorists, it is nearly impossible for the United States to completely refrain from action, as domestic public opinion and the perceived need to maintain credibility internationally necessitate some sort of response.

Jenkins noted that, as in many cases, the United States did not want to get involved in Bosnia, but the Europeans were not capable of putting together a credible response. The United States retains enormous military capacity. Therefore, it is expected to solve the world's problems.

Ayalon said that he understands that America will act in its own self-interest. However, he reminded the workshop that the United States represents more than just Americans and should consider other perspectives. For example, America's decision not to intervene in the Syrian civil war at an early stage caused a huge problem, although the decision could be explained by America's understandable reaction to its disastrous invasion of Iraq. Some of the American participants said that many people in the United States shared his opinion that not acting earlier in Syria had been a mistake. Liepman pointed out, however, that, if you turn the clock back three years, people thought the Syrian civil war would be over quickly. They believed that the president did not have to do anything and the problem would sort itself out on its own.

All the participants agreed that domestic political considerations weigh heavily in decisions to use military force. Politicians often make decisions based on electoral fear rather than sound strategic judgment. This applies to both the United States and Israel. One noted that, in Israel, military action is invariably a political response. Brennan noted that every decision involves partisan politics. President John Kennedy did not want to go to war in Vietnam, but, with the Democrats already having been blamed for the loss of China, he could not be seen to allow more of Asia to fall to communism. After the loss of Cuba, President Lyndon Johnson felt obliged to send American troops into the Dominican Republic in 1965 to prevent what was being portrayed as another Cuba. In 2003, the Democrats were fearful of being seen as weak on Iraq and so went along with the invasion. Later, Obama, who claimed that he left behind a stable Iraq in 2011, felt obliged to intervene to protect his legacy when ISIS forces marched toward Baghdad in 2014. Some of the Israeli participants expressed concern about America's understanding of its own leadership role and its prioritizing issues based on short-term concerns rather than long-term interests.

Events shape policy. Most of the participants agreed that the images of American hostages being beheaded drove American public opinion and demanded action. One government official agreed that atrocities had the ability to change policies, but he warned that the United States and Israel must be careful to distinguish terrorist messaging from real threats to national security. Jenkins pointed out that the beheadings, along with other atrocities, such as the massacre of Yazidis, created tremendous pressure on the United States to intervene, but he added that officials in Washington had themselves contributed to a general hysteria as ISIS forces drove across Iraq. ISIS's successful advance exploited a vacuum as Iraqi forces collapsed and fled. Warnings that ISIS was about to take Baghdad were exaggerated. Hagel said that ISIS "is an imminent threat to every interest we have" and that "this is beyond anything we've seen." General John R. Allen warned Americans that World War III was at hand. Comments like these contributed to the impression that, unless ISIS were destroyed, another 9/11 was inevitable—and the administration would be held responsible.

Deterrence briefly came up again in the discussion. Does deterrence work with nonstate actors? One participant noted that sanctions have no effect on nonstate adversaries. Bombing has no deterrent effect on ISIS. Does that mean there are no means of leverage, only destruction?

This session also returned to the topic of mediafare. There was a great deal of discussion about the need for democratic governments to update their toolkit to address the new challenges posed by adversaries' increasingly effective use of the Internet and social media. Participants affirmed the criticality of messaging and the concept of imagefare in the fight against ISIS and other nonstate adversaries, particu-

[T]he United States and Israel must be careful to distinguish terrorist messaging from real threats to national security.

larly in the context of maintaining public support and international alliances in long-term conflicts.

Noting that it can take days to react to an inflammatory message, photo, or video, Yarchi asked whether govern-

[F]or Israel, the end state is simply to survive.

ments can respond in a more timely manner. The response was that the U.S. government has set up offices to deal specifically with media matters, one of which focuses solely on digital media. Other government observers thought that the U.S. State Department had significantly improved its messaging and countermessaging capabilities in the past several years, adding that the U.S. government has become significantly more adept at responding, especially in native languages. One official noted that there was little messaging could do to prevent the recent beheadings of hostages, and it was difficult to respond. As a whole, however, the U.S. government has become more conscious of and aggressive about messaging and countermessaging. But how do you measure the effectiveness of tweets?

A number of participants highlighted the growing importance of economic issues in counterterrorism, not just efforts to block terrorist financing, but also in the context of threat finance and the adversary's ability to wage economic warfare. They specifically pointed to Israel's vulnerability to sanctions and challenges such as the "Boycott, Divestment, and Sanctions" movement as another area in which Israel's position is more tenuous than that of the United States.

How to define victory proved to be a recurring theme in the workshop. Victory cannot be defined in the traditional sense in asymmetric warfare. David Thaler pointed out that Hezbollah wins by merely not losing, but others recalled that Clausewitz, reflecting on protracted wars in which armies are engaged over long periods of time, wrote that victory is simply about not being exhausted by a prolonged conflict. Some observed that, for Israel, the end state is simply to survive.

CHAPTER TEN

Session Eight: Next Steps

Workshop participants expressed optimism and interest regarding the potential for future collaboration, highlighting the commonality of goals and research ethos. Ayalon said that the workshop proved that the United States and Israel can get together and can bring research from IDI, the Interdisciplinary Center, Haifa University, and Hebrew University to exchange with colleagues at RAND. Although the United States and Israel face different types of terrorist challenges, their experiences are similar enough to warrant mutual reflection and collaborative research. IDI would be forming four working groups to study the four fields outlined at the workshop. Papers would be prepared covering all of these issues.

Jenkins pointed out that, while RAND primarily addresses questions raised in discussions with its clients, it has some capacity to engage in independent research on topics it considers important and that several areas could be identified in which efforts would benefit from useful exchanges like this workshop.

The discussion prompted a number of ideas that would benefit from collaborative efforts:

- **The implications of contemporary asymmetric warfare.** Has the world entered a new area of warfare? Or are these unique challenges faced primarily by Israel and the United States at the moment? If the nature of warfare has fundamentally changed, what are the implications for the United States and Israel?
- **Grand strategy or tailored responses?** Does grand strategy make sense when dealing with the threats posed by specific and very different nonstate adversaries? Although he was quoted several times during the workshop, is Clausewitz relevant to the study of asymmetric conflict?
- **Deterring asymmetric adversaries.** Does deterrence have a role in asymmetric-warfare strategies? If nonstate adversaries feel they have nothing to lose, if they are willing, even eager, to sacrifice their own populations, can deterrence work?
- **Military operations.** There are a number of tactical questions relevant to both the IDF and the U.S. ground forces, such as tunnels, targeted killings, and the use of force.
- **The evolving legal environment.** International law, as it applies to conflict, is changing by custom. More empirical analysis is needed.
- **The impact of protracted warfare on civil liberties and domestic freedoms.** Although not discussed in detail at the workshop, a number of participants, including Ayalon and Jenkins, thought it important to examine the impact of protracted terrorist threats and asymmetric warfare on U.S. and Israeli societies.
- **What is victory?** Decisive victory in warfare is possible only if all people are playing the same game. What does victory mean in asymmetric warfare? Victory is difficult to define

and thus difficult to achieve. Does it mean simply aiming for a better situation after the conflict than before? Does it mean preserving values? The United States and Israel have to aim for a better place. U.S. and Israeli values should be preserved, and the nations must not accede to the terrorist wish that they give them up.

- **What is America's role in the world today?** How broad a range of responsibilities should or can the United States and Israel assume? Can the nations develop an overall grand strategy? The position of the United States is different from that of Israel, which has to focus its international engagement largely on one region.

- **How should the United States deal at once with Europe, the Middle East, and the Indo-Pacific?** Each of these critical regions of the world has its own distinctive political dynamics and security challenges, and each requires its own strategies.

- **How to develop and sustain public support for U.S. international engagement.** This is a critical challenge, particularly given the power of 24/7 media and the fact that adversaries target U.S. public opinion as a key element in their approach to countering U.S. pressures.

- **The changing nature of warfare.** Of particular importance is the proliferation of weapons of mass destruction, especially nuclear weapons.

Afterword

Brian Michael Jenkins

The task of editing the presentations and subsequent discussions provided an opportunity to think further about the issues addressed and to identify some broader themes. Here, then, are some final, personal reflections.

Participants used the terms *asymmetric warfare* and *counterterrorism* synonymously. But one would not call terrorist campaigns by Germany's Red Army Faction or Italy's Red Brigades examples of asymmetric warfare—those actions were terrorism. The introduction of the term *asymmetric warfare* somehow elevates the conflicts above ordinary terrorism.

The concerns of the workshop participants reflect their own particular experience. Israeli frustrations are driven by hybrid groups, such as Hezbollah and Hamas—hybrid because they field well-armed militias and possess large arsenals of rockets but also employ terrorist tactics, while at the same time claiming political authority as a consequence of democratic elections. The concerns of U.S. participants reflect recent American experience—dealing with al Qaeda, the Taliban, and now ISIS and the Islamic State. These are terrorists with territory and not insignificant constituencies, for which they provide social services. They entered Israeli and American consciousness as terrorist groups and now have political pretensions, but others may see them as legitimate political authorities that, at times, use terrorist tactics.

Hezbollah and Hamas are openly dedicated to the destruction of Israel. Al Qaeda has declared war on the United States and has carried out attacks on U.S. soil, and it exhorts its affiliates and followers to continue the campaign, although Jabhat al-Nusra may deviate from this line. ISIS is devoted to armed jihad against the United States; it has killed American citizens, although it has not yet directly attacked the United States. The trappings of statehood—even elections—do not legitimatize armed aggression, let alone terrorism. At most, they make these groups *terrorist states*. But property and claims of state authority do give them a political edge, especially to sympathetic ears, which they effectively exploit.

Asymmetric warfare is hardly a new phenomenon. The assassins sent forth by the Ismaili sect in the 11th century are a splendid example. Unable to mobilize armies that could match the strength of their foes, the Ismailis formulated a strategy of deterrence based on terror. The rulers of states that threatened the Ismaili community, along with their deputies, were targeted for murder by determined assassins who would spend months—even years—planning their operations, infiltrating the staffs of their targets, then dramatically plunging their daggers into the chests of their victims. The willingness of the assassins to forfeit their own lives to carry out their mission, the deliberately dramatic quality of murder, and doing the killing in a

> *The trappings of statehood—even elections—do not legitimatize armed aggression, let alone terrorism.*

public place where witnesses could then spread the word of it created an atmosphere of fear. To attack the Ismailis was to risk assassination. One can have no idea how many missions might have failed. People knew only about the assassins' successes. To be targeted was assumed to mean death.

The Ismailis' assassination strategy worked until the 13th century, when Mongol conquerors in Persia, undeterred by the assassins' threats, destroyed their forts and castles and forced their surrender. Mameluke armies from Egypt, which had by then defeated the Mongols, went on to destroy the remaining Ismaili forts in Syria.

How different are things today? Faced with far superior military forces they cannot possibly hope to match, groups like Hezbollah and Hamas have developed a doctrine of warfare that is unique to their situation. It is a strategy that obviates their enemies' military strength and prowess. These groups ignore the rules of war that place them at a disadvantage but exploit the law to constrain their opponents. They employ terrorist attacks to create alarm and oblige their foes to take extraordinary security measures. They willingly sacrifice their own civilian populations to garner international sympathy and delegitimize their attackers. They excel in political warfare.

The Israeli and American experiences of war elevate the battlefield, where their forces excel. Their opponents elevate the other dimensions of conflict at which they excel. These non-state adversaries understand, perhaps better than their opponents, that war is more than battle. Given the Israelis' own history, it is baffling that they should not easily comprehend that the choice of military objectives must "accentuate the political aspect of [the] struggle," when such thinking was such an integral part of their own struggle to establish the state of Israel. Indeed, the quote comes from the code promulgated by Haganah, Israel's principal resistance organization, in 1946. When creating its formidable military machine after independence, did Israel forget that war is politics?

Today's asymmetric conflicts are asymmetric not only in conventional measures of military strength but also in values. To be sure, both Israel and the United States at times have done things that tarnish their international reputations and invite condemnation—but this does not justify terrorism. Whatever objections one may have to the use of military force in general, or criticisms one may offer about American or Israeli application of military force, or concerns about civilian casualties and collateral damage, there is no moral equivalency between efforts on the one side to constrain military force with rules of engagement and judicial reviews and efforts on the other side to justify any form of violence because the cause claims to be "just" or because of assertions that there is no other way. There is no moral equivalency between efforts to minimize civilian casualties and efforts to deliberately target civilians.

Some of the mediafare tactics of the adversaries work well in an environment of anti-American and anti-Israeli sentiments. This is not a new development. In 1974, the Soviet bloc plus the nonaligned but anti-Western group of countries invited Palestinian leader Yasir Arafat to address the United Nations General Assembly and gave the Palestine Liberation Organization special status in the United Nations. The move was based on hostility to the United States and Israel. The behavior of

> *The Israeli and American experiences of war elevate the battlefield, where their forces excel. Their opponents elevate the other dimensions of conflict at which they excel.*

both nations is not above criticism, and a strong argument can be made that ending the occupation of Palestinian areas might lessen hostility toward Israel in the region and could mute international criticism, but there is little that either country can do to alter the deep underlying hostility, which places both at a disadvantage in the other dimensions of conflict that have become more important.

The United States and Israel are rambunctious democracies, which at times can impede their effectiveness, but they are not the same countries they were when these conflicts began.

The hostility toward Israel and the United States augments antiwar sentiments in Europe and elsewhere. Many do not want to devote resources to costly military establishments they do not believe in and, absent an imminent existential threat, many want to further reduce military spending. Expressions of belief that military power does not work are used to justify such reductions. But, at the same time, they engender hostility toward the two countries that regularly resort to military power.

Faced with the atrocities of ISIS and fears that those who have left their countries to join the group may, in the future, return home to carry out terrorist attacks, a large group of nations, including Arab, as well as European, countries, have joined the United States in a military coalition aimed at destroying the group. Its declaration of an Islamic State has bought it no legitimacy.

Are groups like Hezbollah, Hamas, and the Islamic State each different from one another and all anomalies that reflect unique circumstances in the Middle East? Or do their strategies and tactics reflect the changing nature of war, which, in light of current attitudes and constraints on the application of military force, has become more about the manipulation of perceptions? If these entities and their methods reflect a trend, and, to a certain extent, I think they do, then the search for effective strategies and techniques to counter them are all the more urgent. Such efforts would also have profound implications for the future of armies. War in its more conventional modes may never be obsolete—nations will need armies. But dealing with the growing array of threats will require not merely some additional counterinsurgency or counterterrorism training. It will require some fundamental organizational changes that enable defenders to mobilize and orchestrate resources beyond the military. The United States has done this to a degree on an ad hoc basis, but the results are temporary, regarded as departures, even distractions from the primary missions of armed forces. This was true during the Vietnam War, and it is true today. Analysts have learned much, but the United States and Israel have not yet thought about whether or how to institutionalize what they have learned.

The workshop touched on but did not explore one final issue that concerned its organizers: How will today's open-ended, asymmetrical contests affect the nature of society itself? Will tolerance, privacy, and democracy suffer collateral damage in the quest for security? Concerns were expressed at the workshop, but the experience and expertise of the participants lie elsewhere. The United States and Israel are rambunctious democracies, which can, at times, impede their effectiveness, but they are not the same countries they were when these conflicts began. The United States and Israel claim that democracy is part of their arsenal, but both countries have seen ample examples of public acquiescence to assertions of authority justified by the needs of security and partisan exploitation of fear to advance political agendas—certainly enough to raise concerns. In the meetings and dialogues that follow this workshop, we will have to tackle this issue directly.

Participant Biographies

Israel Democracy Institute

Yohanan (Johan) Plesner is the president of IDI. He grew up in Jerusalem and served as a soldier and officer in an elite IDF commando unit. He graduated magna cum laude with a B.A. in economics from the Amirim Program for Excellence at the Hebrew University of Jerusalem. He also holds an M.P.A. in political economy and international security from the Harvard Kennedy School of Government.

From 1998 to 2000, Plesner served as a senior analyst at UMT Consulting. From 2000 to 2002, he was co-founder and chief executive officer of an international enterprise software company.

In 2005, Prime Minister Ariel Sharon appointed Plesner Head of Special Projects in the Prime Minister's Office. In 2006, he became the first Secretary-General of the new Kadima Party and spearheaded the construction of the party's organizational infrastructure. From 2007 to 2013, he served as a member of the Knesset for Kadima. He was a member of the Constitution, Law and Justice, and Defense and Foreign Affairs Committees, co-chair of the Knesset Lobby for Higher Education, and chairman of the Knesset's permanent delegation to the Council of Europe.

In 2012, Prime Minister Benjamin Netanyahu appointed Plesner to head the Committee for Equality in the Burden of Service, which produced a blueprint for incorporating the ultra-Orthodox into military and national service, one of the most complex social-economic challenges facing the state of Israel.

Jesse Ferris joined IDI in the fall of 2008 as vice president of strategy. His responsibilities include international programming and development.

Ferris was born in the United States but grew up in the Upper Galilee. He served five years in the IDF as a medic, combatant, and team leader, attaining the rank of captain in the reserves. He went on to earn a B.A. with distinction in history from Yale University. Upon graduating from Yale, Ferris co-founded Marketbee Technologies, an enterprise software company based in Jerusalem and San Francisco, where he served as director of marketing and product development until 2002.

Ferris earned his Ph.D. in Near Eastern studies from Princeton University in 2008. His research interests include democracy and foreign policy, international politics of the Middle East, Arab strategic culture, Russian and American foreign policy, and Israeli national security.

Amichay (Ami) Ayalon, a former director of the Israel Security Agency (Shin Bet) and a former commander of Israel's Navy, joined IDI as a senior fellow in December 2012. He has served as a cabinet minister and a member of the Knesset for the Labor party. Along with Sari Nusseibeh, he founded the People's Voice peace initiative in 2002.

Ayalon holds a B.A. in economics and political science from Bar-Ilan University (1980), is a graduate of the U.S. Naval War College in Newport, Rhode Island (1982), holds an M.A. in public administration from Harvard University (1992), and holds an M.A. in law from Bar-Ilan University (2010).

He instituted a code of ethics in the Shin Bet and campaigned for it to be enshrined in law. For these contributions to Israel's security apparatus, the Movement for Quality Government honored him with the Knight of Quality Government award. He is also a recipient of Israel's highest military honor, the Ribbon of Valor, for his part in the raid against the Egyptian fortress at Green Island in 1969.

Ayalon's work at IDI focuses on civil-military relations and issues of national security and democracy. He recently launched a new project on democracies and asymmetrical conflict.

Yuval Shany is the Hersch Lauterpacht Chair in International Law and the dean of the law faculty of the Hebrew University of Jerusalem. He is also a member of the United Nations Human Rights Committee and a Senior Research Fellow at the Israel Democracy Institute. Shany has degrees in law from the Hebrew University (LL.B., 1995, cum laude), New York University (LL.M., 1997), and the University of London (Ph.D., 2001). He has published a large number of books and articles on international courts and arbitration tribunals, international human rights law, international humanitarian law, and international criminal law. He is the recipient of the 2004 American Society of International Law book award (creative legal scholarship) and a 2008 recipient of a European Research Council grant awarded to pioneering research leaders. Shany has taught in a number of law schools in Israel and has been, in recent years, a visiting professor at Harvard University, Columbia University, and Georgetown University and in Sydney, Michigan, Heidelberg, and Amsterdam.

Eli Bahar is a research fellow in IDI's Democratic Principles project and a lecturer at the Hebrew University of Jerusalem and the Tel Aviv University Faculty of Law.

Bahar is a partner and co-founder of the Bahar-Azulay law office, which focuses on civil law, labor law, and security-related cases.

From 1986 to 2011, Bahar served in various capacities with the Israel Security Agency (Shin Bet). In his last position there, he was the head of the legal division and a member of the agency's senior staff. In this position, Bahar was responsible for dealing with legal matters related to operational, constitutional, international, and civic aspects of the agency's activities and worked in close collaboration with the Attorney General, the State Prosecution, and other security agencies.

Bahar received a B.A. in economics and political science, an LL.B. from Tel Aviv University, and an M.A. in social science from Haifa University. He also holds an M.A. in public administration from the Harvard Kennedy School of Government.

Tal Mimran is conducting research as part of IDI's National Security and Democracy project and is the co-author of IDI's *Hebrew Terrorism and Democracy Newsletter.*

Mimran is and was an instructor of international law and property law at several academic institutions in Israel (Hebrew University in Jerusalem, Tel Aviv University, the College of Management, College of Law and Business, Sha'arei Mishpat–The College of Legal Studies). He also mentors Israeli teams for international law competitions, preparing students from the Hebrew University of Jerusalem for the annual Philip C. Jessup Moot Court Competition in Washington, D.C., and students from Tel Aviv University for the National International

Humanitarian Law Competition held by the International Committee of the Red Cross in Israel. He previously coached a team from the College of Law and Business in Ramat Gan in the Willem C. Vis International Commercial Arbitration Moot, held in Vienna, Austria, and served as an arbitrator in that competition.

Mimran interned at the law firm of M. Firon and Co., in the department dealing with antitrust law (competition law) and alternative energy resources. Afterward, he worked for two years as a lawyer in the Department of International Agreements and International Litigation in the Israeli Ministry of Justice, where he dealt mainly with international human rights, international humanitarian law, and international criminal law.

Mimran served as an editor of a human rights journal published under the auspices of the Emile Zola Chair for Human Rights and published articles in that journal. In his reserve duty in the IDF, he serves as a legal adviser in the International Law Department.

Arieh J. Kochavi is a professor of modern European and American history at the University of Haifa. Kochavi currently holds the following offices at the university:

- head, Strochlitz Institute of Holocaust Studies
- head, Herzl Institute for Research and Study of Zionism, University of Haifa
- head, Dr. Reuven Hecht Chair in History
- chair, Editorial Board, Dapim: Studies on the Holocaust
- head, International M.A. Program in Holocaust Studies
- chair, Committee of Ph.D. Studies, Department of History.

Moran Yarchi, Hebrew University of Jerusalem, Israel (2012), is a lecturer at the Sammy Ofer School of Communications, Interdisciplinary Center, Herzliya, Israel. Yarchi's main area of research is political communication, especially the media's coverage of conflicts and public diplomacy. Integrating theories from communication studies, political science and international relations, her studies investigate the ability of political actors to promote their messages through the media.

Robert C. Castel is a research fellow at the University of Haifa's Herzl Institute, with research interests in asymmetric warfare, terrorism, and military innovation. He authored the award-winning study *The Cyberterrorist's Manual for the Internet* (1996, IICC), and his first book, *Innovating for Defeat*, will be published in 2015.

Commissioned in the army, Castel became a paratrooper and a captain in military intelligence. He is a decorated combat veteran, having served in Lebanon and several other areas of operation.

A senior law enforcement official with a background in police special operations, his current position deals with combating wildlife crime.

Yishai Beer is a professor of law at the Radzyner School of Law. He is also a major general (ret.) in the IDF. He is married to Hagit and they have six children.

Beer joined the law faculty at Hebrew University in 1986 and, since then, for 24 years, he has taught courses and seminars in taxation and corporate law. Since 2010, he has lectured at the Interdisciplinary Center. His military career began when he was drafted in 1974 and joined the Paratrooper Brigade. As a young officer, he took part in the 1976 rescue operation in Entebbe. He completed his mandatory service in the IDF in 1978 and continued to serve in the reserves, rising through the ranks and combat commands until commanding a Paratrooper Brigade (as a colonel) in 1995 and the Edom Division (as a brigadier general) in 2000. He also was then commander of the IDF's Brigade Commander's Course. In May 2002, he joined the IDF's general staff when he was appointed president of the Israeli Military Court of Appeals and promoted to major general. From 2007 to 2012, he was a corps (3-5 IDF Army Divisions) commander.

RAND Corporation

Brian Michael Jenkins is a senior adviser to the president of the RAND Corporation and the author of numerous books, reports, and articles on terrorism-related topics, including *Will Terrorists Go Nuclear?* (Prometheus Books, 2008). Jenkins formerly served as chair of the Political Science Department at RAND. On the occasion of the ten-year anniversary of 9/11, he initiated a RAND effort to take stock of America's policy reactions and give thoughtful consideration to future strategy. That effort is presented in *The Long Shadow of 9/11: America's Response to Terrorism* (ed., with John Godges, RAND, 2011).

Commissioned in the infantry, Jenkins became a paratrooper and a captain in the Green Berets. He is a decorated combat veteran, having served in the Seventh Special Forces Group in the Dominican Republic and with the Fifth Special Forces Group in Vietnam. He returned to Vietnam as a member of the Long Range Planning Task Group and received the Department of the Army's highest award for his service.

In 1996, President Clinton appointed Jenkins to the White House Commission on Aviation Safety and Security. From 1999 to 2000, he served as adviser to the National Commission on Terrorism and, in 2000, was appointed to the U.S. Comptroller General's Advisory Board. He is a research associate at the Mineta Transportation Institute, where he directs continuing research on protecting surface transportation against terrorist attacks.

Dalia Dassa Kaye is the director of the Center for Middle East Public Policy and a senior political scientist at the RAND Corporation. In 2011 and 2012, Kaye was a visiting professor and fellow at the University of California, Los Angeles, International Institute and Burkle Center for International Relations. Before joining RAND, Kaye served as a Council on Foreign Relations international affairs fellow at the Dutch Foreign Ministry. She also taught at the University of Amsterdam and was a visiting scholar at the Netherlands Institute of International Relations. From 1998 to 2003, Kaye was an assistant professor of political science and international affairs at the George Washington University. She is the recipient of many awards and fellowships, including a Brookings Institution research fellowship and the John W. Gardner Fellowship for Public Service. Kaye publishes widely on Middle East regional security issues, including in *Survival, Foreign Affairs, The Washington Quarterly, Foreign Policy,* and *Middle East Policy.* She is the author of *Talking to the Enemy: Track Two Diplomacy in the Middle East and South Asia* (RAND, 2007) and *Beyond the Handshake: Multilateral Cooperation in the Arab-Israeli Peace Process, 1991–1996* (Columbia University Press, 2001) and has coauthored a number of RAND monographs on a range of regional security issues. Kaye received her Ph.D. in political science from the University of California, Berkeley.

Rick Brennan is a senior political scientist at the RAND Corporation and a career Army officer with high-level Department of Defense policymaking experience on issues relating to the ongoing war in Iraq, conflict resolution and war termination, homeland defense, and strategic planning. Since 2006, Brennan has served in successive positions in Iraq, including supporting the Multi-National Force–Iraq and U.S. Forces–Iraq as the RAND team leader for the

Joint Interagency Task Force–Iraq and most recently as the senior adviser to the U.S. Forces–Iraq Director of Operations (J3). He has conducted significant research and analysis on special forces operations in Iraq, Phase IV stability and support operations in Iraq between 2003 and 2005, and the transition from military operations to civilian control in Iraq between 2003 and 2012. He has led and participated in several other RAND studies related to homeland security, homeland defense, and military transformation.

Prior to joining RAND, Brennan served as a program manager directing projects related to the revolution in military affairs, information warfare, counterterrorism, and homeland security in support of Office of the Secretary of Defense Director of Net Assessment. He is a former Army officer, who, in addition to infantry command and staff positions, spent six years of his Army career in positions related to national security and defense policy, including two years working within the Office of the Secretary of Defense and three years serving as an assistant professor of international relations at the U.S. Military Academy. A decorated officer, Brennan received the Defense Superior Service Medal, Defense Meritorious Service Medal, Army Meritorious Service Medal (three awards), and the Army Commendation Medal. He holds a Ph.D. in political science and international relations from the University of California, Los Angeles.

Chris Chivvis is an associate director of the International Security and Defense Policy Center and a senior political scientist at the RAND Corporation. He specializes in European and Eurasian security, NATO, military interventions, and deterrence issues. Chivvis is also an adjunct professor at the Johns Hopkins Paul H. Nitze School of Advanced International Studies.

The author of two academic books and several monographs and articles on U.S. foreign and defense policy, Chivvis has worked on Eurasian security and NATO–Russia issues in the Office of the Under Secretary of Defense for Policy. He has also held research positions at the French Institute for International Relations in Paris and at the German Institute for International and Security Affairs in Berlin, and he has taught graduate courses at Johns Hopkins University, New York University, and Sciences Po in Paris.

His work has appeared in *Current History, International Affairs, Journal of Contemporary History, Foreign Policy, National Interest, Survival, Washington Quarterly, International Herald Tribune, Washington Times, Christian Science Monitor, CNN.com,* and other leading publications. Chivvis received his Ph.D. from Johns Hopkins School of Advanced International Studies.

Colin P. Clarke is an associate political scientist at the RAND Corporation, where his research focuses on insurgency and counterinsurgency; unconventional, irregular, and asymmetric warfare (including cyber warfare); and a range of other national and international security issues and challenges. Clarke teaches courses at the University of Pittsburgh and Carnegie Mellon University. In 2011, he spent three months embedded with the Combined Joint Interagency Task Force Shafafiyat in Kabul, Afghanistan, working on anticorruption efforts and analyzing the nexus between terrorists, drug traffickers, and a range of political and economic power brokers. Combined Joint Interagency Task Force Shafafiyat was commanded by LTG H. R. McMaster. Clarke is in the process of completing *Terrorism, Inc.: The Financing of Terrorism, Insurgency, and Irregular Warfare,* to be published in 2015 by Praeger Security International. His book features case studies of Hamas, Hezbollah, al Qaeda, and the Islamic State.

James Dobbins is a senior fellow and distinguished chair in diplomacy and security at the RAND Corporation. He most recently served as director of the RAND International

Security and Defense Policy Center. Dobbins has held State Department and White House posts, including Assistant Secretary of State for Europe, Special Assistant to the President, Special Adviser to the President and Secretary of State for the Balkans, and Ambassador to the European Community. Dobbins has served on numerous crisis-management and diplomatic troubleshooting assignments as special envoy for Afghanistan and Pakistan, Kosovo, Bosnia, Haiti, and Somalia for the administrations of Barack Obama, George W. Bush, and Bill Clinton. Diplomatic assignments include the withdrawal of American forces from Somalia, the American-led multilateral intervention in Haiti, the stabilization and reconstruction of Bosnia, and the NATO intervention in Kosovo. In the wake of September 11, 2001, he was named as the Bush administration's representative to the Afghan opposition, with the task of putting together and installing a broadly based successor to the Taliban regime. He represented the United States at the Bonn Conference that established the new Afghan government, and, on December 16, 2001, he raised the flag over the newly reopened U.S. embassy.

Shira Efron is a Ph.D. candidate at the Pardee RAND Graduate School and an assistant policy analyst at the RAND Corporation. She has an M.A. in international relations from New York University and a B.S. in biology and computer science from Tel Aviv University. At RAND, Efron is working on problems related to food security; water; energy; the Middle East; intelligence, surveillance, and reconnaissance; and education. Her Ph.D. dissertation explores the feasibility of employing unmanned aerial systems for agriculture in Africa.

Prior to coming to RAND, Efron was primarily conducting research and analysis on Middle East geopolitical issues. She was the policy director and country representative of the Institute for Inclusive Security in Israel. Before that, she was a policy analyst at the Center for American Progress, where she edited the *Middle East Bulletin*, a multiweekly online publication for a high-level U.S. government and stakeholder audience focusing on the intersection of U.S. political, economic, and security interests in the Middle East, North Africa, and South Asia. She is a member of several nonpartisan organizations that work to promote a two-state solution to the Israeli-Palestinian conflict while ensuring Israel's security. Efron was also a research analyst at a hedge fund in New York, an editor at the Israeli newspaper *Ha'aretz*, and a reporter in the IDF.

Her interests include Middle East geopolitics; food security; water; energy; homeland security; terrorism; defense policy; education; science, technology, engineering, and mathematics; and ethnic conflicts.

Todd C. Helmus is a senior behavioral scientist who specializes in irregular warfare, counterterrorism, and security cooperation. Helmus has authored numerous studies that focus on improving U.S. efforts to counter militant recruitment and decrease popular support for terrorism and insurgency. Since 2010, Helmus has worked closely with U.S. Special Operations forces in Afghanistan, where he served as an adviser to U.S. commanders and led studies on U.S. efforts to train Afghan Special Security forces. In 2008, he also served in Baghdad as an adviser to Multi-National Force–Iraq. Helmus is the lead author of *Enlisting Madison Avenue: The Marketing Approach to Earning Popular Support in Theaters of Operation* (with Christopher Paul and Russell W. Glenn, RAND, 2007). He received his Ph.D. in clinical psychology from Wayne State University.

George Jameson served more than 30 years in the CIA and the U.S. Intelligence Community as a senior counsel, as director of the CIA's policy and coordination office, and as a manager of legislative affairs at the CIA and the office of the Director of National Intelligence. His responsibilities included reviewing the legality and propriety of operations, including covert action and counterterrorism, war crimes issues, foreign relationships, information and privacy, and intelligence community policy and reform. A private consultant, Jameson is an adjunct staff member at the RAND corporation and chairman of the nonprofit Council on Intelligence Issues. He has authored works, including *Intelligence and the Law* (American Bar Association, 2011), *Fixing Leaks* (with James Bruce, RAND, 2013), and "The Law of Counterterrorism, What Next?" (in Lynne Zusman, ed., *The Fundamentals of Counterterrorism Law*, American Bar Association, 2014). He earned an A.B. from Harvard College and a J.D. from the Marshall-Wythe School of Law at the College of William and Mary.

David E. Johnson is a senior political scientist at the RAND Corporation and a retired Army colonel whose research focuses on military doctrine, history, innovation, civil-military relations, and professional military education. From June 2012 to July 2014, Johnson was on loan to the Army to establish and direct the Chief of Staff of the Army Strategic Studies Group for GEN Raymond Odierno. He is the author of *Fast Tanks and Heavy Bombers: Innovation in the U.S. Army, 1917–1945* (Cornell University Press, 1998); *Learning Large Lessons: The Evolving Roles of Ground Power and Air Power in the Post–Cold War Era* (RAND, 2007); and *Hard Fighting: Israel in Lebanon and Gaza* (RAND, 2011); and he is co-author of *The 2008 Battle of Sadr City: Reimagining Urban Combat* (RAND, 2013).

Patrick Johnston is an associate political scientist at the RAND Corporation. He specializes in counterinsurgency and counterterrorism, with a particular focus on Afghanistan and the Philippines. His articles have been published or are forthcoming in a range of peer-reviewed journals, including *American Economic Review, International Security, Security Studies*, and *Civil Wars*. Before coming to RAND, Johnston was a fellow at the Harvard Kennedy School of Government, Stanford University, and the United States Institute of Peace. He completed his Ph.D. in political science at Northwestern University in 2009.

Eric V. Larson is a senior policy researcher at the RAND Corporation and a professor at the Pardee RAND Graduate School. His most recent research has included analysis of the ground-force requirements for weapons of mass destruction–elimination operations; Chinese and Indian defense spending; al Qaeda strategy, ideology, propaganda, and discourse; public support for terrorism and insurgency; influence activities, strategic communication, and information support operations; irregular warfare; American and foreign attitudes toward U.S. military operations; media and public opinion; and social science approaches to national security analysis. Larson has recently been working on a historical review of U.S. defense planning since 2001, assured access in the Pacific, women in Special Operations forces, and the development of social science–based intelligence analysis tradecraft. Prior to joining RAND, he worked as a policy and systems analyst at the National Security Council and as a research staff member at the Institute for Defense Analyses. Larson received his A.B. in political science from the University of Michigan and his M.Phil. and Ph.D. in policy analysis from the Pardee RAND Graduate School.

Andrew Liepman is a senior policy analyst at the RAND Corporation. He retired in August 2012 as the principal deputy director of the National Counterterrorism Center after a career of more than 30 years in the CIA. Liepman spent much of his career on Middle East and terrorism issues. He served for three years at the Department of State and in a variety of Intelligence Community assignments, including positions in the Nonproliferation Center and the National Intelligence Council. He was the deputy chief of the CIA's Office of Near East and South Asian Analysis; the deputy director of the Office of Weapons Intelligence, Arms Control, and Nonproliferation; the chief of the Office of Iraq Analysis; and the deputy chief of

CIA's Counterterrorism Center. Immediately prior to joining RAND, he served in the Office of the Director of National Intelligence, first as the deputy director of the National Counterterrorism Center for intelligence and finally as the principal deputy. Liepman earned a B.S. from the University of California, Berkeley.

Richard H. Solomon is a senior fellow at the RAND Corporation. He previously served as president of the United States Institute of Peace from 1993 to 2012.

Solomon was assistant secretary of state for East Asian and Pacific affairs from 1989 to 1992. He negotiated the Cambodia peace treaty and the first United Nations "Permanent Five" peacemaking agreement; had a leading role in the dialogue on nuclear issues between the United States and South and North Korea; helped establish the Asia-Pacific Economic Cooperation initiative; and led U.S. negotiations with Japan, Mongolia, and Vietnam on important bilateral matters. In 1992 and 1993, Solomon served as U.S. ambassador to the Philippines, where he coordinated the closure of the U.S. naval bases and developed a new framework for bilateral and regional security cooperation.

Solomon previously served as director of policy planning at the Department of State and as a senior staff member of the National Security Council. In 1995, Solomon was awarded the State Department's Foreign Affairs Award for Public Service, and he has received awards for policy initiatives from the governments of Korea and Thailand. In 2005, he received the American Political Science Association's Hubert H. Humphrey career award for "notable public service by a political scientist."

Solomon began his career as a professor of political science at the University of Michigan and served as head of the Political Science Department at RAND. He holds a Ph.D. in political science, with a specialization in Chinese politics, from the Massachusetts Institute of Technology.

Michael Spirtas is the associate director of the Defense and Political Sciences Department and a senior political scientist at the RAND Corporation. He has worked in a number of policy areas, including defense force modernization, defense planning scenarios, command and control of joint task forces, U.S. Air Force force structure, lessons from Middle East conflicts, coalition military operations, and strategy in Afghanistan and Pakistan. Spirtas has also worked on force development issues for the Office of the Under Secretary of Defense for Policy and as a special assistant to the Director for Operational Plans and Joint Matters, U.S. Air Force. He received his Ph.D. in political science (international relations) from Columbia University.

David Thaler is a senior defense analyst at the RAND Corporation. He has led or participated in studies related to building partner capacity; foreign policy and military implications of developments in Iran, Iraq, and the broader Middle East and Iranian domestic politics; military capabilities for counterinsurgency and counterterrorism (irregular warfare); and the strategies-to-tasks framework for force planning. He was detailed from RAND to the Air Staff from 1993 to 1995 and again from 2003 to 2004, and he supported U.S. Air Force efforts to prepare for and execute the Quadrennial Defense Reviews. He has also conducted research on first-strike stability, aircraft maintenance, and military readiness. Thaler received his M.I.A. (master of international affairs) in international security policy and the Middle East from Columbia University.

Invited Guests

- David Lubarsky
- Allan Myer
- William (Bill) Recker
- U.S. government officials

Workshop Agenda

This appendix reproduces the workshop agenda.

RAND CENTER FOR MIDDLE EAST PUBLIC POLICY

International Programs at RAND

THE ISRAEL
DEMOCRACY
INSTITUTE

DEMOCRACIES FACING ASYMMETRIC CONFLICTS

A Collaborative Workshop between the
Israel Democracy Institute and the RAND Corporation

December 2-4, 2014
Washington, D.C.
RAND Corporation · 1200 South Hayes Street · Arlington, VA 22202 · 703-413-1100

Tuesday, December 2
    19:00–21:00   **Opening Dinner, Bonefish Grill, Pentagon Row**
                         (Group can meet at Ritz lobby at 18:45 to walk to Bonefish Grill,
                         1101 S Joyce Street B-26, Arlington, VA 22202, 703-412-2837)
                         Welcome remarks by Brian Michael Jenkins and Admiral Amichay Ayalon

Wednesday, December 3
    8:00-8:30   **Breakfast** at RAND

    8:30–10:00   Session One:
                         Introduction: The changing terrorist threat and America's evolving
                         response
                         Presentation by Brian Michael Jenkins
                         Discussant: Prof. Yishai Beer

    10:00–10:30   **Break**

    10:30–12:00   Session Two:
                         Distinctions in asymmetric warfare
                         Presentation by Admiral Amichay Ayalon
                         Discussant: Andrew Liepman

    12:00–13:00   **Working Lunch**

    13:00–14:30   Session Three:
                         Relevant RAND research on counter-insurgency and counter-terrorism
                         Presentations by Colin Clarke, Patrick Johnston, and Eric Larson

    14:30–16:00   Session Four:
                         From warfare to mediafare
                         Presentation by Dr. Moran Yarchi
                         Discussant: Todd Helmus

16:00–16:30	**Break**
16:30–17:30	**Session Five:** Lawfare Presentation by Adv. Eli Bahar Discussant: W. George Jameson
17:30–18:00	**Wrap Up Discussion**
18:30–21:00	**Dinner, Ritz-Carlton Hotel, Diplomat Room** *(Begin with drinks at 18:30, and dinner at 19:00)*

Thursday, December 4

8:30-9:00	**Breakfast** at RAND
9:00–10:30	**Session Six:** Political and diplomatic dimensions: lessons from around the globe Discussion led by James Dobbins
10:30–11:00	**Break**
11:00–12:30	**Session Seven:** Review of challenges, objectives, and initial findings with invited guests from the government Discussion led by Admiral Amichay Ayalon and Brian Michael Jenkins
12:30–13:30	**Working Lunch**
13:30–15:00	**Session Eight:** Next steps
15:00	**Workshop Concludes**

Thursday, December 4

17:45–19:45	**Public Event at the National Press Club (Optional)** *(529 14th St. NW, 13th Floor, Washington, DC 20045, 202-662-7500)* Discussion with Admiral Amichay Ayalon and Brian Michael Jenkins Moderated by James Kitfield, Senior Fellow, Center for the Study of the Presidency and Congress; contributing editor, National Journal

References

Ayalon, Ami, Elad Popovich, and Moran Yarchi, "From Warfare to Imagefare: How States Should Manage Asymmetric Conflicts with Extensive Media Coverage," *Terrorism and Political Violence*, 2014.

Bahney, Benjamin, Howard J. Shatz, Carroll Ganier, Renny McPherson, and Barbara Sude, *An Economic Analysis of the Financial Record of al-Qa'ida in Iraq*, Santa Monica, Calif.: RAND Corporation, MG-1026-OSD, 2010. As of July 7, 2015:
http://www.rand.org/pubs/monographs/MG1026.html

Brennan, Rick, Jr., Charles P. Ries, Larry Hanauer, Ben Connable, Terrence K. Kelly, Michael J. McNerney, Stephanie Young, Jason H. Campbell, and K. Scott McMahon, *Ending the U.S. War in Iraq: The Final Transition, Operational Maneuver, and Disestablishment of United States Forces—Iraq*, Santa Monica, Calif.: RAND Corporation, RR-232-USFI, 2013. As of July 7, 2015:
http://www.rand.org/pubs/research_reports/RR232.html

Brother 'Imad, "Income and Outcome Report for Al-Anbar Province in Iraq 2," Combating Terrorism Center at West Point, Harmony Database document NMEC-2007-633700, 2007a. As of July 7, 2015:
https://www.ctc.usma.edu/posts/income-and-outcome-report-for-al-anbar-province-in-iraq-2-original-language

———, "Income and Outcome Report for Al-Anbar Province in Iraq 4," Combating Terrorism Center at West Point, Harmony Database document NMEC-2007-633919, 2007b. As of July 7, 2015:
https://www.ctc.usma.edu/posts/income-and-outcome-report-for-al-anbar-province-in-iraq-4-original-language

Churchill, Winston, *A Roving Commission: My Early Life*, New York: Charles Scribner's Sons, 1930.

Clausewitz, Carl von, *On War*, Princeton, N.J.: Princeton University Press, 1976.

Connable, Ben, and Martin C. Libicki, *How Insurgencies End*, Santa Monica, Calif.: RAND Corporation, MG-965-MCIA, 2010. As of July 7, 2015:
http://www.rand.org/pubs/monographs/MG965.html

Hallin, Daniel C., *The "Uncensored War": The Media and Vietnam*, Berkeley, Calif.: University of California Press, 1989.

Johnston, Patrick B., *Countering ISIL's Financing*, Santa Monica, Calif.: RAND Corporation, CT-419, 2014. As of July 7, 2015:
http://www.rand.org/pubs/testimonies/CT419.html

Johnston, Patrick B., and Benjamin Bahney, "Hitting ISIS Where It Hurts: Disrupting ISIS's Cash Flow in Iraq," *New York Times*, August 13, 2014.

Jones, Seth G., and Martin C. Libicki, *How Terrorist Groups End: Lessons for Countering al Qa'ida*, Santa Monica, Calif.: RAND Corporation, MG-741-1-RC, 2008. As of July 7, 2015:
http://www.rand.org/pubs/monographs/MG741-1.html

Leites, Nathan Constantin, and Charles Wolf, Jr., *Rebellion and Authority: An Analytic Essay on Insurgent Conflicts*, Santa Monica, Calif.: RAND Corporation, R-462-ARPA, 1970. As of July 7, 2015:
http://www.rand.org/pubs/reports/R0462.html

Mutz, Diana C., "The Great Divide: Campaign Media in the American Mind," *Daedalus*, Vol. 141, No. 4, Fall 2012, pp. 1–15.

Paul, Christopher, Colin P. Clarke, and Beth Grill, *Victory Has a Thousand Fathers: Sources of Success in Counterinsurgency*, Santa Monica, Calif.: RAND Corporation, MG-964-OSD, 2010. As of July 7, 2015: http://www.rand.org/pubs/monographs/MG964.html

Paul, Christopher, Colin P. Clarke, Beth Grill, and Molly Dunigan, *Paths to Victory: Lessons from Modern Insurgencies*, Santa Monica, Calif.: RAND Corporation, RR-291/1-OSD, 2013. As of July 7, 2015: http://www.rand.org/pubs/research_reports/RR291z1.html

Prior, Markus, "Media and Political Polarization," *Annual Review of Political Science*, Vol. 16, 2013, pp. 101–127.

Public Law 104-132, Antiterrorism and Effective Death Penalty Act of 1996, April 24, 1996. As of July 7, 2015: http://www.gpo.gov/fdsys/pkg/PLAW-104publ132/content-detail.html

Reagan, Ronald, "Combatting Terrorism," Washington, D.C.: White House, National Security Decision Directive 138, April 3, 1984. As of July 7, 2015: http://fas.org/irp/offdocs/nsdd/nsdd-138.pdf

"The Uses of Military Power," remarks prepared for delivery by the Hon. Caspar W. Weinberger, Secretary of Defense, to the National Press Club, Washington, D.C., November 28, 1984. As of July 7, 2015: http://www.pbs.org/wgbh/pages/frontline/shows/military/force/weinberger.html

Transparency International, "Corruption Perceptions Index 2009," November 17, 2009. As of July 7, 2015: http://www.transparency.org/research/cpi/cpi_2009

U.S. Department of State, *Patterns of Global Terrorism: 1987*, August 1988.

Vreese, Claes H. de, and Hajo G. Boomgaarden, "Media Message Flows and Interpersonal Communication: The Conditional Nature of Effects on Public Opinion," *Communication Research*, Vol. 33, No. 1, February 2006, pp. 19–37.

Waltz, Kenneth N., *Man, the State, and War: A Theoretical Analysis*, New York: Columbia University Press, 1959.

———, *Theory of International Politics*, New York: McGraw-Hill, 1979.

Zaller, John, *The Nature and Origins of Mass Opinion*, Cambridge, UK: Cambridge University Press, 1992.

———, *A Theory of Media Politics: How the Interests of Politicians, Journalists, and Citizens Shape the News*, draft, October 24, 1999.